Power Control Circuits

Newnes Circuits Manual Series

Audio IC Circuits Manual   R.M. Marston
CMOS Circuits Manual   R.M. Marston
Diode, Transistor and FET Circuits Manual   R.M. Marston
Electronic Alarm Circuits Manual   R.M. Marston
Op-amp Circuits Manual   R.M. Marston
Optoelectronics Circuits Manual   R.M. Marston
Timer/Generator Circuits Manual   R.M. Marston

Power Control
Circuits Manual

R.M. Marston

NEWNES

An imprint of Butterworth-Heinemann Ltd

To Brenda, the true power in the Marston household

Butterworth-Heinemann Ltd
Linacre House, Jordan Hill, Oxford OX2 8DP

PART OF REED INTERNATIONAL BOOKS

OXFORD   LONDON   BOSTON
MUNICH   NEW DEl HI   SINGAPORE   SYDNEY
TOKYO   TORONTO   WELLINGTON

First published 1990
Reprinted 1992

British Library Cataloguing in Publication Data
Marston, R.M. (Raymond Michael), *1937*
   Power control circuits manual
   1. Electronic equipment. Circuits
   I. Title
   621.3815'3

ISBN 0 7506 0690 8

Printed and bound in Great Britain
by Billing & Sons Ltd, Worcester

Contents

Preface

Electronic power control circuits can be used to manually or automatically control the brilliance of lamps, the speed of motors, the temperature of heating devices such as electric fires or radiators, or the loudness of audio signals, etc. This control can be achieved using electromechanical switches or relays, or electronic components such as transistors, SCRs, TRIACs, or power ICs, etc. This book takes an in-depth look at the whole subject of electronic power control, and in the process presents the reader with a vast range of useful circuits and diagrams.

The book is specifically aimed at the practical design engineer, technician, and experimenter, as well as the electronics student and amateur, and deals with its subject in an easy-to-read, down-to-earth, non-mathematical but very comprehensive manner. Each chapter deals with a specific aspect of power control, and starts off by explaining the basic principles of its subject and then goes on to present the reader with a wide range of practical application circuits.

The book is split into eight distinct chapters. Chapter 1 explains the basic principles of electrical-electronic power control, and Chapter 2 shows practical control circuits using conventional switches and relays. Chapter 3 describes ways of using CMOS devices as low-power electronic switches, and Chapters 4 and 5 deal with AC and DC power control systems and present sixty-eight practical application circuits.

One of the most important sections of the book is Chapter 6, which takes a close look at ways of controlling DC motors, including those of the 'stepper' and servo type, and presents forty-two practical circuits. The final two chapters deal with audio power control and DC power supply systems, and present a further total of sixty-two circuits.

Throughout the volume, great emphasis is placed on practical 'user' information and circuitry, and the book abounds with useful circuits and graphs; a total of 283 diagrams are included. Most of the solid state devices used in the practical circuits are modestly priced and readily available types, with universally recognized type numbers.

1 Basic principles

An electrical or electronic power control circuit can be defined as any circuit that is used to control the distribution or the levels of AC or DC power sources. Such circuits can be used to control (either manually or automatically) the brilliance of lamps, the speed of motors, the temperature of heating devices such as electric fires or radiators, or the loudness of audio signals, etc., or they can be used to manually switch power to these or other devices, or to switch power automatically when parameters such as temperature or light intensity, etc., go beyond pre-set limits.

A variety of devices can be used in power control applications. These range from simple switches and electromechanical devices such as relays and solenoids, which can be used as low-speed power switches, to solid-state devices such as transistors, FETs, CMOS multiplexers, SCRs or TRIACs, or power ICs, etc., which can be used as high-speed power switches or magnitude controllers. This opening chapter describes basic electronic power control principles, and shows how the above devices can be used in power control applications at levels ranging from a fraction of a milliwatt to several kilowatts.

Power switching circuits

All electric power controllers fit into either of the two basic categories: power switchers (such as a lamp on/off switch); and power level controllers (such as a lamp dimmer). *Figure 1.1* shows examples of three basic types of power switching circuit, and *Figures 1.2* to *1.5* illustrate the operating principles of four different types of power level control circuit.

1

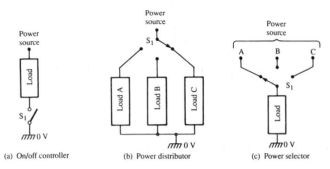

Figure 1.1 *Three basic types of power switching circuit*

The three basic switching circuit types are the on/off controller (*Figure 1.1(a)*), which is used to switch power to a single load, the power distributor (*Figure 1.1(b)*), which switches power to one or other of several different loads, and the power selector (*Figure 1.1(c)*), which feeds one or other of several different power sources to a single load.

Note in *Figure 1.1* that power switching is shown via ordinary electric switches, but in practice these can easily be replaced by sets of relay contacts or by any of a variety of types of solid-state switch.

DC power control

Figure 1.2 shows the basic circuit of a simple DC power level controller in which 0 to 12 V is available on RV_1 slider and is fed to the load via a current-boosting voltage follower buffer stage. Note that this type of circuit is not very efficient, since all unwanted power is 'lost' across the buffer stage. If, for example, the load is

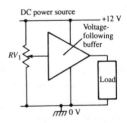

Figure 1.2 *Simple DC power level controller*

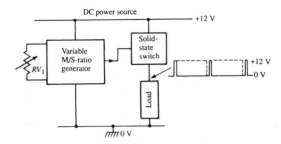

Figure 1.3 *Switched-mode DC power level controller*

fed with 1 V and draws 1 A (thus consuming 1 W), 11 V are lost in the 1 A buffer, which thus consumes 11 W, so the circuit operates with an efficiency of only 8.33 per cent.

Figure 1.3 shows a different type of power level controller, which operates with an efficiency of about 95 per cent. Here, power is fed to the load via a solid-state power switch that is activated via a square-wave generator with a variable mark-space (M/S) ratio or duty cycle. For explanatory purposes, assume that the duty cycle is variable from 5 to 95 per cent via RV_1, and that the solid-state switch is 100 per cent efficient. In this case the circuit operates as follows.

When the solid-state switch is open, 0 V are generated across the load, and when it is closed the full 12 V supply line voltage is generated across the load. When the switch is activated via the variable M/S-ratio generator, the mean voltage of the load (integrated over one duty cycle) is proportional to the generator's duty cycle.

Thus, if the generator has a 50 per cent duty cycle (i.e., a 1:1 M/S-ratio, or equal on and off times), the mean load voltage equals 50 per cent of the 12 V supply value, or 6 V. Similarly, if the duty cycle is 5 per cent, the mean load voltage is 600 mV, and if the duty cycle is 95 per cent the mean voltage is 11.4 V. Since power consumption is proportional to the square of the mean supply voltage, it can be seen that this circuit enables load power to be varied from 0.25 to 90.25 per cent of maximum via RV_1.

Note that in practice a peak of only 200 mV or so is usually lost across a solid-state power switch, so this type of circuit operates with a typical efficiency of about 95 per cent at all times, and is widely used in DC lamp-brilliance and motor-speed control applications.

AC power control

Figures 1.4 and *1.5* show two ways of adapting the switched-mode variable-duty-cycle power control technique for use in AC applications. The *Figure 1.4* circuit uses a 'phase-triggered' switching technique widely used for controlling the AC power feed to filament lamps, etc., which have fairly long thermal. time con-

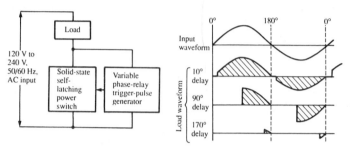

Figure 1.4   *Variable phase-delay-switching AC power controller, with waveforms*

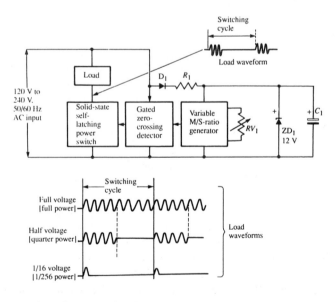

Figure 1.5   *Burst-fire (integral-cycle) AC power controller*

stants, and to electric power drills and motors, which have high mechanical inertia, and the *Figure 1.5* circuit uses a 'burst fire' technique widely used for controlling electric fires, etc., which consume high currents and have long thermal time constants.

In *Figure 1.4*, power is applied to the load via a self-latching solid-state power switch (a triac, etc.), which can be triggered (via a variable phase-delay network and a trigger pulse generator) at any point during each power half-cycle, but which automatically unlatches again at the end of each half-cycle as the AC voltage falls briefly to zero. The diagram shows the load voltage waveforms that can be generated.

Thus, if the power switch is triggered just after the start of each half cycle (with near-zero phase delay), the mean load voltage equals almost the full supply value, and the load consumes near-maximum power. If the switch is triggered half way through each half cycle (with 90° phase delay), the mean load voltage equals half the supply value, and the load thus consumes one quarter of maximum power. Finally, if the switch is triggered near the end of each half cycle (with near-180° phase delay), the mean load voltage is near-zero, and the load consumes minimal power.

The *Figure 1.4* phase-triggered power control technique has several advantages. It is highly efficient (typically, better than 95 per cent), enables the load power to be fully varied over a wide range, and, since switching occurs at the power line frequency, enables lamp brilliance to be varied without flicker. A major disadvantage is that, since power may be switched abruptly from zero to a high-peak value (particularly at about 90° delay), the resulting high current surges can generate substantial RFI (radio frequency interference). This type of circuit is thus not suitable for feeding high-current loads such as electric fires, etc.

Burst-fire control

High-current loads such as electric fires can be efficiently power-controlled without generating significant RFI by using the *Figure 1.5* burst-fire technique, in which power bursts of complete half-cycles are fed to the load at regular line-frequency-related intervals. Thus, if bursts are repeated at 8-cycle intervals, the mean load voltage equals the full supply line value if the bursts are of 8-cycle duration, or half voltage (equals quarter power) at 4-cycle duration, or one sixteenth voltage (equals $\frac{1}{256}$ power) at one half-cycle duration, etc.

The burst-fire technique generates near-zero RFI because it switches power to the load only very near the start of line half-cycles, when the instantaneous line voltage (and load current) is very low. This is achieved by using a line-driven zero-crossing detector, which is gated via a variable M/S-ratio generator and gives an output only if gated on and the line voltage is below 7 V or so. The detector output triggers the self-latching solid-state device (triac) used to switch power to the load. The variable M/S-ratio generator is powered from a 12 V DC supply derived from the AC power line via D_1–R_1 and ZD_1–C_1.

The burst-fire or 'integral cycle power control' technique is highly efficient, but enables the load's power consumption to be varied only in a number of discrete half-cycle steps. When driving electric heaters, however, this last-mentioned factor is of little importance, and the system can easily be used to give precision automatic room-temperature control with the aid of suitable temperature-sensing thermistors or thermostats, etc.

Electric switch basics

The simplest type of power control device is the ordinary electric switch, which comes in several basic versions; a selection of these are depicted by the symbols of *Figure 1.6.*. The simplest switch is

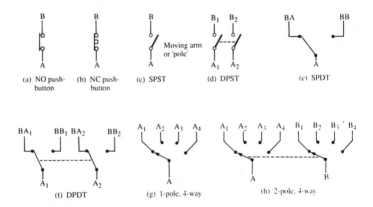

Figure 1.6 *Some basic switch configurations*

the push-button type, in which a spring-loaded conductor can be moved so that it does or does not bridge (short) a pair of fixed contacts. These switches come in either normally-open (NO) form (*Figure 1.6(a)*), in which the button is pressed to short the contacts, or in normally-closed (NC) form (*Figure 1.6(b)*), in which the button is pressed to open the contacts.

The most widely used switch is the moving arm type, which is shown in its simplest form in *Figure 1.6(c)*) and has a single spring-loaded (biased) metal arm or 'pole' that has permanent electrical contact with terminal A but either has or has not got contact with terminal B, thus giving an on/off switching action between these terminals. This type of switch is known as a single-pole single-throw (SPST) switch.

Figure 1.6(d) shows two SPST switches mounted in a single case with their poles 'ganged' together so that they move in unison to make a double-pole single-throw (DPST) switch.

Figure 1.6(e) shows a single-pole double-throw (SPDT) switch in which the pole can be 'thrown' so that it connects terminal A to either terminal BA or BB, thus enabling the A terminal to be coupled in either of two different directions or 'ways'.

Figure 1.6(f) shows a ganged double-pole double-throw (DPDT) version of the above switch. Note that these multiway switches can be used in either simple on/off or multiway power distribution/ selection applications.

Figure 1.6(g) shows a switch in which the A terminal can be coupled to any of four others, thus giving a '1-pole, 4-way' action. Finally, *Figure 1.6(h)* shows a ganged 2-pole version of the same switch. In practice, switches can be designed to give any desired number of poles and ways.

Two other widely used electric switches are the pressure-pad type, which takes the form of a thin pad easily hidden under a carpet or mat and which is activated by body weight, and the microswitch, which is a toggle switch activated via slight pressure on a button or lever on its side, thus enabling the switch to be activated by the action of opening or closing a door or window or moving a piece of machinery, etc.

Electro mechanical relay basics

The conventional electromagnetic relay is really an electrically operated switch, and is a very useful power-control device.

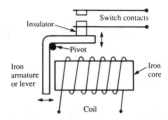

Figure 1.7   *Basic design of standard electromagnetic relay*

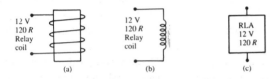

Figure 1.8   *Alternative ways of representing a 12 V, 120R relay coil*

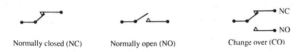

Figure 1.9   *The three basic types of contact arrangement*

Figure 1.7 illustrates its operating principle. Here, a multiturn coil is wound on an iron core to form an electromagnet that can move an iron lever or armature which in turn can close or open one or more sets of switch contacts. Thus, the operating coil and the switch contacts are electrically fully isolated from one another, and can be shown as separate elements in circuit diagrams.

The main characteristics of the relay coil are its operating voltage and resistance values, and *Figure 1.8* shows different ways of representing a 12 V, 120 Ω coil. The symbol of *Figure 1.8(c)* is the easiest to draw, and carries all vital information. Practical relays may have coils designed to operate from a mere few volts DC up to the fall AC power line voltage, etc.

There are three possible basic types of relay contact arrangement, these being normally closed (NC), normally open (NO), and changeover (CO), as shown in *Figure 1.9*. Practical relays often carry more than one set of contacts, with all sets ganged. Thus, the term 'DPCO' simply means that the relay carries two sets of changeover contacts. Actual contacts may have electrical rating up to several hundred volts, and up to tens of amperes.

Relay configurations

Figures 1.10 to *1.13* show basic ways of using ordinary relays. In *Figure 1.10*, the relay is wired in the non-latching mode, in which push-button switch S_1 is wired in series with the relay coil and its supply rails, and the relay closes only while S_1 is closed.

Figure 1.11 shows the relay wired to give self-latching operation. Here, NO relay contacts RLA/1 are wired in parallel with activating switch S_1. RLA is normally off, but turns on as soon as S_1 is closed, making contacts RLA/1 close and lock RLA on even if S_1 is subsequently re-opened. Once the relay has locked on it can be turned off again by briefly breaking the supply connections to the relay coil.

Note in the above circuits that the relay can be switched in the AND mode via several series-wired switches, so that the relay turns on only when all switches are closed, or in the OR mode via several

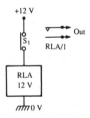

Figure 1.10 *Non-latching relay switch*

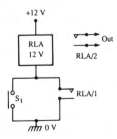

Figure 1.11 *Self-latching relay switch*

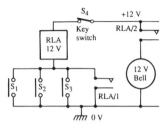

Figure 1.12   *Simple burglar alarm*

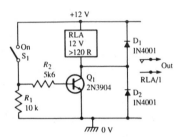

Figure 1.13   *Transistor driven relay with two-diode coil damper*

parallel-wired switches, so that the relay turns on when any of these switches are closed. *Figure 1.12* shows these modes used in a simple burglar alarm, in which the relay turns on and self-latches (via RLA/1) and activates an alarm bell (via RLA/2) when any of the S_1 to S_3 switches are briefly closed (by opening a door or window or treading on a mat, etc.). The alarm can be enabled or turned off via key switch S_4.

Some relay coils can be activated via only a few volts and milliamperes, enabling them to be turned on and off via simple transistor (or IC) circuitry if desired, as shown in the example of *Figure 1.13*, where a coil current of 100 mA can be obtained via an S_1 current of less than 4 mA. Note that relay coils are highly inductive and can generate back-e.m.f.s of hundreds of volts if their coil currents are suddenly broken; these back-e.m.f.s can easily damage electronic coil-driver circuitry. This danger can be over-come by connecting protective 'damping' diodes D_1 and D_2 to the coil as shown. D_1 prevents the RLA-Q_1 junction from swinging more than 600 mV above the positive supply rail value, and D_2 stops it from swinging more than 600 mV below the zero-volt rail value.

Reed relay basics

Another type of electromechanical relay is the 'reed' type, which consists of a springy pair of opposite-polarity magnetic reeds with gold- or silver-plated contacts sealed into a glass tube filled with protective gases, as shown in *Figure 1.14*. The opposing magnetic fields of the reeds normally hold their contacts apart, so they act as an NO switch, but these fields can be effectively cancelled or reversed (so that the switch closes) by placing the reeds within an externally-generated magnetic field, which can be derived from either an electric coil that surrounds the glass tube, as shown in *Figure 1.15*, or by a permanent magnet placed within a few millimetres of the tube, as shown in *Figure 1.16*.

Practical reed relays are available on both NO and CO versions, and their contacts can usually handle maximum currents of only a few hundred milliamperes. Coil-driven types can be used in the same way as normal relays, but typically have a drive-current sensitivity ten times better than a standard relay.

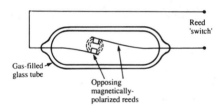

Figure 1.14 *Basic structure of reed relay*

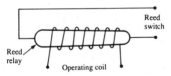

Figure 1.15 *Reed relay operated by coil*

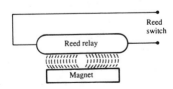

Figure 1.16 *Reed relay operated by magnet*

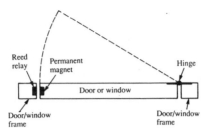

Figure 1.17   *Method of using a reed relay/magnet combination to give burglar protection to a door or window*

A major advantage of the reed relay is that it can be 'remote activated' at a range of several millimetres via an external magnet, enabling it to be used in many home-security applications; *Figure 1.17* illustrates the basic principle. The reed relay is embedded in a door or window frame, and the activating magnet is embedded adjacent to it in the actual door or window so that the relay changes state whenever the door/window is opened or closed. Several of these relays can be interconnected and used to activate a suitable alarm circuit, if desired.

Transistor devices

So far we have looked only at electromechanical power control devices. Many solid-state devices are also used in power control applications, and the simplest of these is the discrete bipolar transistor, which is usually used in the switching mode as shown in *Figures 1.18* and *1.19*. In the case of the npn transistor the switch

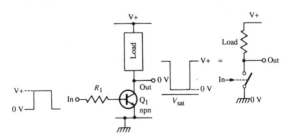

Figure 1.18   *Bipolar npn transistor switch circuit*

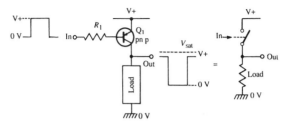

Figure 1.19 *Bipolar pnp transistor switch circuit*

load is wired between Q_1 collector and supply positive, and in the case of the pnp device it is wired between Q_1 collector and the 0 V rail. In both cases the switch driving signal is applied to Q_1 base via R_1, which has a typical resistance about twenty times greater than the load resistance value.

In the npn circuit Q_1 is cut off (acting like an open switch), with its output at the positive supply voltage value, with zero input signal applied, but can be driven to saturation (so that it acts like a closed switch and passes current from collector to emitter) by applying a large positive input voltage, under which condition the output equals Q_1's saturation voltage value (typically 200 mV to 600 mV).

The action of the pnp circuit (*Figure 1.19*) is the reverse of that described above, and Q_1 is driven to saturation (with its output a few hundred millivolts below the supply voltage value) and passes current from emitter to collector with zero input drive voltage applied, and is cut off (with its output at zero volts) when the input equals the positive supply rail value.

A useful variation of the transistor switch is the optocoupler, which comprises an infra-red LED (light-emitting diode) and matching phototransistor, mounted close together (optically coupled) in a light-excluding package, as shown in the basic application circuit of *Figure 1.20*. Here, when SW_1 is open zero current flows in

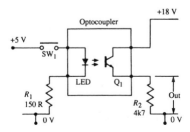

Figure 1.20 *Basic optocoupler switching circuit*

the LED, so Q_1 is in darkness and also passes zero current (it acts like an open switch), so zero output appears on R_2. When SW_1 is closed, current flows in the LED via R_1, illuminating Q_1 and making it act as a closed switch that generates an R_2 output voltage which can thus be controlled via the R_1 input current, even though R_1 and R_2 are fully isolated electrically. In practice, this device can be used to optocouple either digital (switching) or analogue signals, and offers hundreds of volts of isolation between the input and output circuits.

FET devices

A field-effect transistor (FET) is a 3-terminal voltage-controlled current generator with a near-infinite input or GATE impedence. FETs are available in two basic versions, giving either an *enhancement* or a *depletion* mode of operation. Enhancement mode FETs pass zero current when the gate voltage is zero, and the current rises (is enhanced) when the gate is forward biased. Depletion mode FETs give the reverse of this action; they pass a maximum current when the gate voltage is zero, and the current falls (depletes) when the gate is forward biased. In this book we will consider enhancement mode FETs only.

FETs are available under several exotic names related to details of their construction. Thus, names such as IGFET (insulated gate FET), MOSFET (metal oxide silicon FET) and JFET (junction FET) are widely used among low-power devices, and names like VMOS, HMOS and HEXFET are common among high-power devices.

FETs are available as n-channel and p-channel types, which are analogous to npn and pnp transistors respectively. *Figure 1.21* and

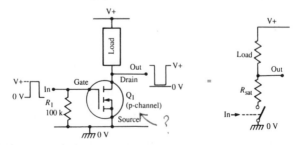

Figure 1.21   *Enhancement-mode n-channel MOSFET swtiching circuit*

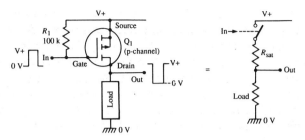

Figure 1.22 *Enhancement-mode p-channel MOSFET switching circuit*

1.22 show how to use n-channel and p-channel enhancement-mode MOSFETs as simple electronic switches. The n-channel MOSFET acts as an open switch when its gate voltage is zero, and as a closed switch (in series with a 'saturation resistance') which passes current from *drain* to *source* when its gate is at the positive supply rail value. The p-channel MOSFET gives the reverse action, and acts as a closed switch (plus saturation resistance) which passes current from *source* to *drain* when its gate voltage is zero, and as an open switch when its gate is at the positive supply rail value.

The 'closed switch' saturation resistance of a MOSFET may vary from a few hundred ohms in a low-power device to a fraction of an ohm in a high-power one, and gives a voltage-divider action with the load resistor that determines the closed-state output voltage of the circuit. Thus, if the *Figure 1.21* circuit has a 10 V supply and a 900 Ω load, it will give a 'closed' output of 1V0 at an R_{SAT} value of 100 Ω, or 10 mV at an R_{SAT} value of 0.9 Ω.

Note that a MOSFET's gate terminal has a near-infinite input resistance, and if allowed to 'float' can accumulate electrostatic charges that can destroy the device. Some MOSFETs have a built-in zener diode to give protection against the danger, as shown in *Figure 1.23*.

Figure 1.23 *Symbol of MOSFET n-channel device with internal zener diode gate protection*

Basic CMOS switches

One of the most important digital IC families is that known as CMOS, and this whole family is based on the simple complementary *MOS*FET digital inverter circuit of *Figure 1.24*, which com-

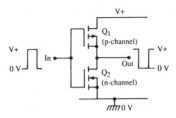

Figure 1.24   *Basic CMOS digital inverter circuit*

prises nothing more than a p-channel and an n-channel enhancement-mode MOSFET wired in series between the two supply lines, with the two gates tied together at the input terminal and with the output taken from the junction of the two devices. In use, the input is at either zero volts (logic-0) or full positive supply rail voltage (logic-1).

Figure 1.25(a) shows the digital equivalent of this circuit with a logic-0 input, with Q_1 acting as a closed switch in series with an R_{SAT} of 400 Ω, and Q_2 acting as an open switch. The circuit thus draws zero quiescent current but can 'source' fairly large drive

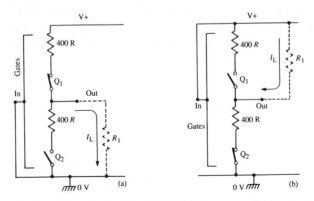

Figure 1.25   *Equivalent circuits of the CMOS digital inverter with (a) logic-0 and (b) logic-1 inputs*

currents into an external output-to-ground load via the 400 Ω ouput resistance (R_1) of the inverter.

Figure 1.25(b) shows the equivalent of the inverter circuit with a logic-1 input. In this case Q_1 acts as an open switch, but Q_2 acts as a closed switch in series with an R_{SAT} of 400 Ω. This circuit thus draws zero quiescent current, but can 'sink' fairly large currents via an external supply-to-output load via its internal 400 Ω output resistance (R_2).

Thus, the basic CMOS digital inverter has a near-infinite input impedance, draws near-zero (typically 0.01 μA) quiescent current with a logic-0 or logic-1 input, can source or sink substantial output currents, and has an output that is inherently short-circuit proof via its 400 Ω output impedence. It gives an excellent low-power switching action.

CMOS bilateral switches

One very important member of the CMOS family is the bilateral switch or transmission gate, which is shown in basic and symbolic forms in *Figure 1.26(a)* and *(b)*. The importance of this device is that (like a normal switch or set of relay contacts) it can conduct current in either direction (bilaterally), whereas a single transistor or FET can conduct in one direction only (from collector to emitter in the case of an npn device, or from emitter to collector in a pnp device).

The operation of the *Figure 1.26(a)* circuit is fairly simple. The device comprises an n-channel and a p-channel MOSFET wired in inverse parallel (drain-to-source and source-to-drain), but with their gate signals applied in antiphase via a pair of CMOS inverter

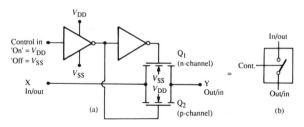

Figure 1.26 *(a) Basic circuit and (b) symbol of simple CMOS bilateral switch or transmission gate*

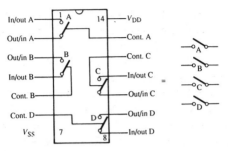

Figure 1.27   *The 4016B and 4066B quad bilateral switches each act as four independent SPST switches*

stages to give the bilateral switching action. Thus, when the control signal is at logic-0 the gate of Q_2 is driven to logic-1 and that of Q_1 is driven to logic-0, and under this condition both MOSFETs act as open switches between the circuit's X and Y points. When, on the other hand, the control signal is at logic-1 the gate of Q_2 is at logic-0 and that of Q_1 is at logic-1, so both MOSFETs act as closed switches, and a low resistance (equal to the R_{SAT} value) exists between the X and Y points of the circuit.

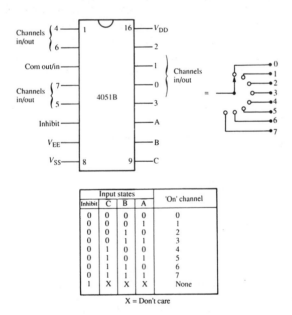

Input states				'On' channel
Inhibit	C	B	A	
0	0	0	0	0
0	0	0	1	1
0	0	1	0	2
0	0	1	1	3
0	1	0	0	4
0	1	0	1	5
0	1	1	0	6
0	1	1	1	7
1	X	X	X	None

X = Don't care

Figure 1.28   *The 4051B acts as a single-pole 8-way bilateral switch*

Note when the control input is at logic-1 that signal currents can flow in either direction between the X and Y terminals (via Q_1 in one direction, or Q_2 in the other), provided that the signals are within the logic-level voltage limits; the X and Y terminals can thus be used as either in or out terminal.

Practical CMOS bilateral switch circuits are usually a bit more complex than shown in *Figure 1.26*, and give typical on resistances of only 100 Ω or so. All practical CMOS ICs of these types house several bilateral switches. The 4016B and 4066B ICs, for example, each house four switches configured as independent SPST switches, as shown in *Figure 1.27*, and the 4051B houses a set of switches and a logic network configured to act as a single-pole eight-way bilateral switch, as shown in *Figure 1.28*.

Silicon controlled rectifier basics

The solid-state power control devices examined so far are useful with low- to medium-voltage DC supplies. Silicon controlled rectifiers (SCRs) and triacs, by contrast, are solid-state power control devices intended for use with medium- to high-voltage AC supplies. *Figure 1.29* shows the symbol and basic 'power switch' application circuit of an SCR. Note in this and other circuits that follow that alternative component values are given for use with 120 V and 240 V power lines, the 240 V values being noted in parentheses.

The SCR can (as its name and symbol imply) be regarded as a 3-terminal silicon rectifier that can be controlled via its GATE terminal. Normally, it acts as an open switch, but if its *anode* is

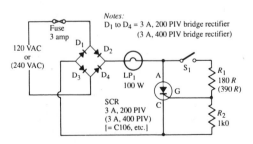

Figure 1.29 *Full-wave on-off SCR circuit with DC power load*

positive to its *cathode* it can be made to switch on and act as a forward biased rectifier by applying a brief trigger current to its gate terminal. If its resulting anode-to-cathode current exceeds a minimum 'holding' value (usually a few milliamperes) the SCR self-latches into the 'on' state and stays there until its anode-to-cathode current falls below the minimum holding value, at which point the SCR reverts to the open-switch state. Thus, the *Figure 1.29* circuit acts as follows.

The AC power line signal is full-wave rectified via D_1–D_4 and converted into a waveform that goes from zero to maximum and then back to zero again in each AC half-cycle; this waveform as applied to the SCR anode via lamp-load LP_1. Thus, if S_1 is open zero gate drive is fed to the SCR, which acts like an open switch, and the lamp is off. If S_1 is closed, SCR gate drive is applied via R_1–R_2, so just after the start of each half-cycle the SCR turns on and self-latches until the end of the half-cycle, at which point it automatically turns off again as its forward current falls below the minimum holding value. This process repeats in each half-cycle, and the lamp thus operates at almost full power under this condition.

The SCR anode falls to only a few hundred millivolts when the SCR turns on, so note that S_1 and R_1–R_2 consume little mean power, but S_1 can be used (via the SCR) to control very large power loads. Also note that the lamp load is shown placed on the DC side of the bridge rectifier, and this circuit is thus shown as for use with DC loads; it can be modified for use with AC loads by simply placing the load on the AC side of the bridge, as in *Figure 1.30*.

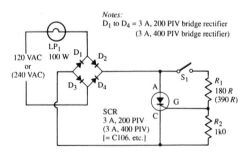

Figure 1.30   *Full-wave on-off SCR circuit with AC power load*

Triac basics

The SCR is a unidirectional device which can conduct current only from anode to cathode. A triac, on the other hand, is a bidirectional device that can conduct current in either direction between its two main terminals (MT_1 and MT_2), and it can thus be used to directly control AC power. *Figure 1.31* shows the standard symbol of the triac, and *Figure 1.32* shows it used as a simple AC power switch that can be used to replace the *Figure 1.30* SCR-based design.

The triac can, for most practical purposes, be regarded as a pair of SCRs wired in inverse parallel (and thus able to conduct MT_2 currents in either direction) and sharing a common gate terminal. The device action is such that it can be triggered by either positive or negative gate currents, irrespective of the polarity of the MT_2 current, and it thus has four possible triggering modes or 'quadrants', signified as follows:

I+ mode = MT_2 current positive, gate current positive.
I− mode = MT_2 current positive, gate current negative.
III+ mode = MT_2 current negative, gate current positive.
III− mode = MT_2 current negative, gate current negative.

Most triac data sheets carry information relating to the device sensitivity in each of the four quadrants. The trigger current sensitivity is always greatest when the MT_2 and gate currents are both of the same polarity (either both positive or both negative), and is usually about half as great when they are of opposite polarity.

The operation of the *Figure 1.32* circuit is thus quite simple. When S_1 is open, the triac acts as an open switch and the lamp passes zero current, but when S_1 is closed the triac is gated on via R_1 and self-latches shortly after the start of each half-cycle, thus switching full power to the lamp load.

Note that SCRs and triacs can be used to apply variable power to

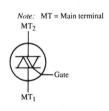

Note: MT = Main terminal
MT_2

Gate

MT_1

Figure 1.31 *Triac symbol*

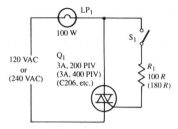

Figure 1.32 *Simple triac AC power switch circuit*

DC or AC loads by using either the burst-fire or the phase-triggered techniques described in *Figure 1.4* and *1.5*.

SCR/triac 'rate-effect'

An SCR or triac is turned on by feeding a trigger signal to its gate. Unfortunately, internal capacitances inevitably exist between the anode and gate of an SCR or the MT_1 terminal and gate of a triac, and if a sharply rising voltage is fed to an SCR/triac anode/MT_1 terminal it can cause enough gate-voltage breakthrough to trigger the SCR/triac on. This unwanted 'rate-effect' turn-on can be caused by supply line transients, and sometimes occurs at the moment that supplies are switch-connected to the SCR/triac; the problem is particularly severe when driving inductive loads such as electric motors, in which load currents and voltages are out of phase.

Rate-effect problems can usually be overcome by wiring an $R–C$ 'snubber' network between the anode and cathode of an SCR or MT_1 and MT_2 of a triac, to limit the voltage rate-of-rise to a safe value, as shown, for example, in the triac power switch circuit of *Figure 1.33*, where $R_2–C_1$ form the snubber network.

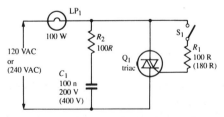

Figure 1.33 *Simple triac power switch circuit with $R_2–C_1$ 'snubber' network to give rate-effect suppression*

RFI suppression

Each time a resistance-driving AC-powered SCR or triac is gated on its load current switches sharply (in a few microseconds) from zero to a value determined by its load resistance and supply voltage values. This switching action inevitably generates a pulse of RFI, which is least when the device is triggered close to the 0° and 180° 'zero crossing' points of the supply line waveform (at which the switch-on currents are at their minimum), and is greatest when the device is triggered 90° after the start of each half cycle (where the switch-on currents are at their greatest).

The RFI signal magnitude is also proportional to the cable length linking the SCR/triac to its power load. Note that the RFI pulses repeat at double the supply line rate if the SCR/triac is triggered in each supply line half-cycle, and that RFI generation can thus be particularly annoying in simple lamp dimmer circuits. Fortunately, such problems can usually be eliminated by fitting the dimmer with a simple L-C RFI-suppression network, as shown in *Figure 1.34*.

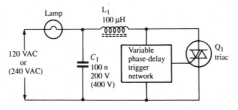

Figure 1.34 *Basic lamp dimmer circuit with RFI suppression via C_1–L_1*

Here, the L-C filter is fitted close to the triac, and greatly reduces the rate-of-rise of supply line transition currents, thus eliminating RFI.

Diacs and quadracs

The SCR and triac are the two best-known members of a family of solid-state thyristor 'trigger' devices. Two other important members of this family are the diac and the quadrac.

Figure 1.35 shows the standard circuit symbol of the diac, a 2-

Figure 1.35 *Diac symbol*

terminal bilateral trigger device that can be used with voltages of either polarity. Its basic action is such that, when connected across a voltage source via a current-limiting load resistor, it acts like a high impedance until the applied voltage rises to about 35 V, at which point the diac triggers and acts like a 30 V zener diode, so 30 V are developed across the diac and the remaining 5 V are developed across the load resistor. The diac remains in this state until its forward current falls below a minimum holding value (this occurs when the supply voltage is reduced below the 30 V 'zener' value), at which point the diac turns off again.

The diac is most often used as a trigger device in phase-triggered triac variable power control applications, as in the basic lamp dimmer circuit of *Figure 1.36*. Here, in each power line half-cycle, the R_1–C_1 network applies a variable phase-delayed version of the half-cycle to the triac gate via the diac, and when the C_1 voltage rises to 35 V the diac fires and delivers a 5 V trigger pulse (from C_1) into the triac gate, thus turning the triac on and simultaneously applying power to the lamp load and removing the drive from the R–C network. The mean power to the load (integrating over a full half-cycle period) is thus fully variable from near-zero to maximum via R_1.

Some triacs are manufactured with a built-in diac in series with the triac gate; such devices are known as quadracs, and use the circuit symbol shown in *Figure 1.37*.

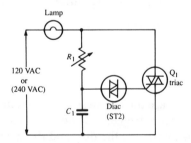

Figure 1.36 *Basic diac-type variable phase-delay lamp dimmer circuit*

Figure 1.37 *Quadrac symbol*

The unijunction transistor

The unijunction transistor (UJT) is a 3-terminal trigger device that is often used to trigger SCRs or triacs. *Figure 1.38(a)* shows its

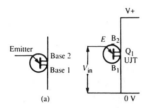

Figure 1.38 *(a) Circuit symbol and (b) basic usage circuit of the UJT*

circuit symbol. It is normally used in the basic way shown in *Figure 1.38(b)*, with its B_2 terminal taken to the positive supply rail and its B_1 terminal grounded, and with an input signal voltage applied to its emitter (E) terminal; the action of this circuit is as follows.

Normally, when the input (E) voltage is very low, the UJT has a near-infinite E-to-B_1 input impedence. If the input voltage is slowly increased a point is reached, at a 'peak-point' voltage (V_P) of about 60 per cent of the supply value, at which the input impedance starts to fall and the input starts to draw a trigger current; if this current is allowed to exceed a minimum 'peak-point emitter current' (I_P) value of a few microamperes, the UJT enters a regenerative switching phase in which the input impedance drops to a mere 20 Ω or so. Once triggered, the UJT input impedance remains low until the input current is reduced below a 'valley-point' value (I_v) of a few milliamperes, at which point the input impedance starts to switch high again.

UJT oscillator

The UJT is usually used in a relaxation oscillator circuit shown in basic form in *Figure 1.39*. Here, the UJT input is taken from the

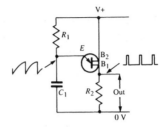

Figure 1.39   *Basic UJT relaxation oscillator circuit*

C_1–R_1 timer network, and the output is taken from R_2, which is wired between B_1 and ground. This circuit operates as follows.

When power is first applied C_1 is discharged, so E is at 0 V and has a near-infinite input impedance. As soon as power is applied, C_1 starts to charge via R_1 and generates an exponentially rising voltage on E; eventually, after a period determined by C_1–R_1, the E voltage reaches the UJT's peak-point value, and the input impedance starts to switch low. Under this condition C_1 acts as a low impedance power source, and the E-to-B_1 impedance thus switches to a very low value, making C_1 rapidly discharge into R_2 and thus generating a brief but powerful output pulse until, eventually, the E input current falls below the UJT's valley-point value, at which point the emitter reverts to the high impedance state, and C_1 starts to recharge via R_1. The process repeats ad infinitum, with an exponential sawtooth waveform being developed at the C_1-R_1 junction and a pulse waveform being developed across R_2.

Note that for this circuit to work correctly the R_1 value must be large enough to limit currents to less than the UJT's I_v value of a few milliamperes (i.e., greater than a few kilohms, but small enough to allow currents to exceed the UJT's minimum I_P current of a few microamperes (i.e., less than 500 kΩ or so). Typically, R_1 can have any value in the range 4k7 to 500 kΩ, enabling pulse delay times to be varied over a wide range via R_1 and making the circuit suitable for use in a variety of phase-delayed power control applications

Isolated-input switching

A major attraction of the UJT oscillator is that it can generate high-current output pulses (up to several hundred milliamperes) while consuming a fairly low mean current (a couple of milliamperes). One important application of the UJT oscillator is shown in the isolated-input AC power switch circuit of *Figure 1.40*, where the

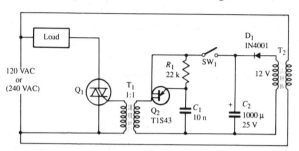

Figure 1.40 *UJT-triggered isolated-input AC power switch*

oscillator is DC power via T_2–D_1–C_2 and SW_1 and operates at several kilohertz, and thus delivers roughly fifty trigger pulses to the triac gate (via isolation pulse transformer T_1) during each AC power line half-cycle. Consequently, the triac is triggered by the first pulse occurring in each power half-cycle, and this appears within a few degrees of the start of the half-cycle.

The triac is thus turned on almost permanently when SW_1 is closed, and virtually full power is applied to the AC load. The trigger circuit is, however, fully isolated from the AC power source via T_1 and T_2, and can be turned on and off by switching a few milliamperes of DC current via SW_1. In practice, SW_1 can easily be replaced by any type of electronic sensing circuitry, which can be fully isolated from the AC power line.

2 Switch and relay circuits

In Chapter 1 we took a brief look at electromechanical switches and relays, explaining their basic operating principles and showing a few simple applications. In this chapter we expand on this theme, and show a large selection of useful circuits.

Lamp switching circuits

The simplest power control circuit is that used to turn a filament lamp on and off. In AC mains-powered applications this takes the *Figure 2.1* series form, with SW_1 connected to the *live*, *phase* or '*hot*' power line and the lamp wired to the *neutral* or *safe* line, to minimize the consumer's chances of getting a shock when changing lamps; this simple circuit allows the lamp to be switched on and off from one point only.

Figure 2.2 shows how to switch a lamp from either of two points, by using a two-way switch at each point, with two wires (known as strapping wires) connected to each switch so that one or other wire carries the current when the lamp is turned on.

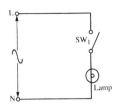

Figure 2.1   *Single-switch on/off AC lamp control circuit*

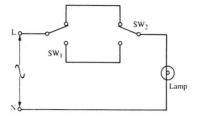

Figure 2.2 *Two-switch on/off AC lamp control circuit*

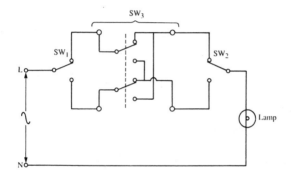

Figure 2.3 *3-switch on/off AC lamp control circuit*

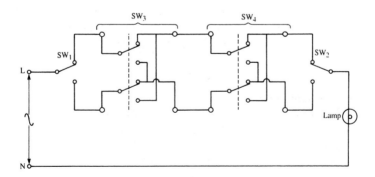

Figure 2.4 *4-switch on/off AC lamp control circuit*

Figure 2.3 shows the above circuit modified to give lamp switching from any of three points. Here, a ganged pair of 2-way switches (SW$_3$) are inserted in series with the two strapping wires, so that the SW$_1$–SW$_2$ lamp current flows directly along one strapping wire path when SW$_3$ is in one position, but crosses from one strapping wire path to the other when SW$_3$ is in the alternative position.

Note that SW$_3$ has opposing pairs of output terminals shorted together. In the electric wiring industry such switches are available with these terminals shorted internally, and with only four terminals externally available (as indicated by the small white circles in the diagram); these switches are known in the trade as 'intermediate' switches.

In practice, the basic *Figure 2.3* circuit can be switched from any desired number of positions by simply inserting an intermediate switch into the strapping wires at each desired new switching position. *Figure 2.4*, for example, shows the circuit modified for four position switching.

Multi-input switching circuits

Note that although the *Figure 2.4* circuit is electrically simple it may be difficult to install physically, since each intermediate switch needs at least a 3-core (live, neutral, and ground) heavy duty cable feeding into it and a similar cable feeding out, and these must be buried in channels cut into plaster and masonary if concealed wiring is wanted.

These installation problems can be minimized in any of three basic ways. One technique often used in long corridors in hotels and guesthouses, etc., where lights need to be switched from many different points, is to use simple push-operated timer switches which stay closed for only a minute or so once operated, connected in parallel as in *Figure 2.5*, thus enabling the switch wiring to be greatly simplified.

Another technique is to use light-duty multiway switching circuitry (with wiring that can easily be hidden in plaster, etc.) to activate a low-voltage relay (powered from a mains-derived 12 V DC supply), which then switches the AC-powered lamp via one set of relay contacts, as shown, for example, in the 2-way switching circuit of *Figure 2.6*.

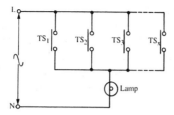

Figure 2.5 *Multi-input AC lamp switching circuit using simple timer switches*

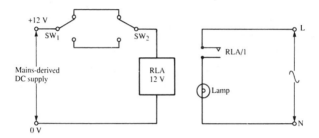

Figure 2.6 *Multi-input relay-activated AC lamp switching circuit*

Finally, the best technique of all is to switch the lamp via a relay that is electronically activated via multi-input push-button control circuitry, thus enabling very simply light-duty switch wiring to be used. Practical circuits of this type are described later in this chapter (see *Figure 2.19*).

Relay configurations

The simplest way of using a relay is in the basic non-latching mode, with the activating switch in series with its coil as already shown in *Figure 1.10*, so that the relay closes only when the switch is closed.

In practice, a relay may be activated via several switches, either wired in series to give AND-logic operation, or in parallel to give OR-logic operation, as shown in *Figures 2.7* and *2.8*; in the AND-logic circuit the relay turns on only when all series-connected switches (SW_1 AND SW_2 AND SW_3, etc.) are closed at the same time; in the OR-logic circuit the relay turns on when any of the

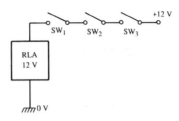

Figure 2.7   *AND logic switching*

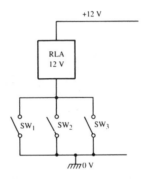

Figure 2.8   *OR logic switching*

parallel-connected switches (SW$_1$ OR SW$_2$ OR SW$_3$, etc.) are closed.

Coil damping

Relay coils are inductive, and may generate back-e.m.f.s of hundreds of volts if their coil currents are suddenly broken. These back-e.m.f.s can easily damage switch contacts or solid-state devices connected to the coil, and it is thus often necessary to 'damp' these coil back-e.m.f.s via protective diodes. *Figure 2.9* and *2.10* show examples of such circuits.

In *Figure 2.9* the coil damping is provided via D$_1$, which prevents switch-off back-e.m.f.s from driving the RLA–SW$_1$ junction more than 600 mV above the positive supply rail value. This form of protection is adequate for many practical applications.

In *Figure 2.10* the damping is provided via two diodes that stop the RLA–SW$_1$ junction from swinging more than 600 mV above the

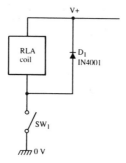

Figure 2.9 *Single-diode coil damper*

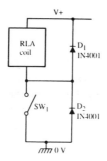

Figure 2.10 *Two-diode coil damper*

positive supply rail or below the zero-volt rail. This form of protection is adequate for even the most critical needs, and is recommended for all applications in which SW_1 is replaced by a transistor or other 'solid state' switch.

High-Z relay switches

Normal and reed-type relays are low-impedance devices, with typical coil impedances of only a few tens or hundreds of ohms. Their effective input impedances can easily be increased, however (so that they act as high-impedance or 'high-Z' relay switches), by simply interposing a transistor or IC driver stage between the relay coil and the input signal. *Figures 2.11* to *2.14* show four variations of this theme.

In *Figure 2.11*, Q_1 is wired as a common-emitter amplifier and

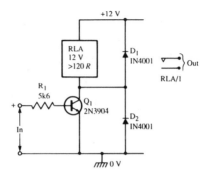

Figure 2.11   *Non-latching transistor-driven relay switch*

increases the effective relay coil-current sensitivity by a factor of about ×100 (the current gain of Q_1) and also increases the voltage sensitivity to a few volts. R_1 limits the Q_1 input current to a safe value and also dictates the effective input impedance of the circuit (R_1 plus roughly 1k0). D_1 and D_2 damp the coil back-e.m.f.s. The RLA/1 contacts can control external circuitry.

The *Figure 2.11* circuit gives a non-latching action in which RLA is off when the input is below 600 mV and is on when the input exceeds a few volts. The circuit can be made to give a self-latching action by modifying it as in *Figure 2.12*, where NO contacts RLA/2 (plus NC switch SW_1) are in parallel with Q_1 and thus bypass Q_1 and self-latch RLA once it has been initially activated. Once it has self-latched, the relay can only be turned off again by either opening SW_1 or breaking the supply-line connections.

The above two circuits have input impedances of a few kilohms; if desired, impedance of 10 MΩ or more can be obtained by driving

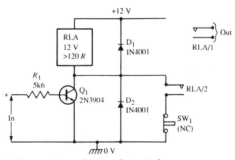

Figure 2.12   *Self-latching transistor-driven relay switch*

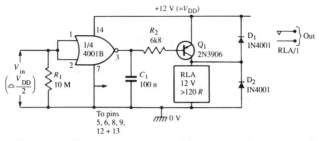

Figure 2.13 *Simple high-impedance relay switch*

Q_1 via a CMOS buffer made from one or more stages of a 4001B quad 2-input NOR gate wired in the inverter mode (by shorting the gate input pins together), as shown in *Figures 2.13* and *2.14*.

In *Figure 2.13* only a single CMOS inverter stage is used. Consequently, to ensure that the relay turns off when the input voltage is low, the Q_1 driver must be a pnp device. In *Figure 2.14*, two CMOS inverter stages are wired in series, to give zero overall signal inversion, and Q_1 can thus by an npn device.

Note in both the above circuits that the input impedance equals the R_1 value, and that the relay turns on (or off) when the input signal goes above (or falls below) the half-supply-voltage (approximately) 'transition' voltage value of the CMOS input gate, at which value the gate operates in the linear mode; C_1 inhibits any high frequency or transient instability that may occur in the gate when in this linear mode.

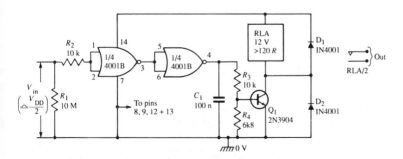

Figure 2.14 *Modified high-impedance relay switch*

A simple burglar alarm

Figure 2.15 shows a simple but useful burglar alarm that consumes a mean current of only 1 μA when in the 'standby' mode (with SW_1 and all series-connected SW_2 sensor switches closed). When any of the SW_2 sensor switches open, a 'high' voltage is fed to the input of the 4001B CMOS gate, which is wired as an inverting buffer and thus drives relay RLA on via Q_1. As RLA turns on it self-latches via contacts RLA/2 and activates the alarm bell via contacts RLA/1. Note that R_2–R_3–C_1 act as a transient-suppressing filter network that protects the alarm against false activation via lightening strikes, etc.

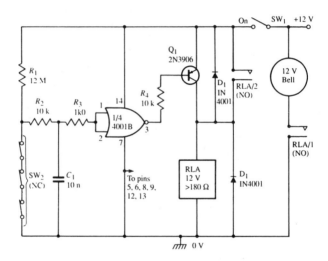

Figure 2.15　*Simple burglar alarm, activated by series-connected NC switches*

Bistable circuits

A relay can be made to give a bistable action in which it turns on and self-latches when a SET push-button switch is operated, and can then subsequently be turned off again only by operating a RESET push-button switch, by using the connections shown in *Figure 2.16*. Here, two NOR gates (taken from a 4001B CMOS IC)

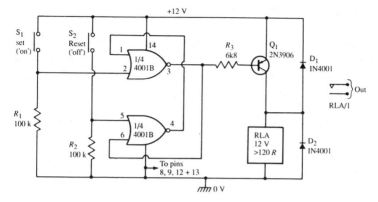

Figure 2.16 *Simple bistable relay switch*

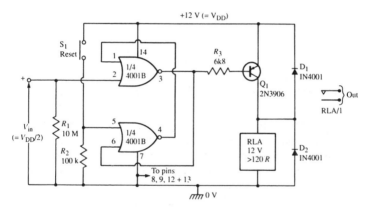

Figure 2.17 *Self-latching high-impedance relay switch*

are wired as a simple manually-activated bistable multivibrator that has one of its outputs taken to the relay coil via the Q_1 common-emitter buffer stage.

The *Figure 2.16* circuit actually changes state as its set or reset input signal rises through the half-supply-voltage 'transition' value of the CMOS gate, and this fact enables the circuit to be modified to give a self-latching relay action as in *Figure 2.17*. Here, the relay turns on and self-latches when the input voltage rises above the 'transition' value, and the relay can subsequently be turned off only by removing (or reducing) the input voltage and pressing the reset switch. This circuit has an input impedance of 10 MΩ.

Figure 1.18 shows a very useful type of push-button operated relay switch. It can use a single control switch to turn the relay on and off, but any number of such switches can in fact be wired in parallel, enabling the relay to be remote-controlled from any number of input points. The circuit action is such that the relay changes state each time an input switch is operated (pressed and released): thus, if the relay is initially on, it will turn off when any switch is next operated, and on again when any switch is operated after that, and so on. The circuit thus gives a 'binary' relay switching action.

The *Figure 2.18* circuit is designed around a 4013B CMOS dual D-type flip-flop, with one flip-flop disabled by taking its inputs to ground and the other configured as a divide-by-two circuit (by shorting its not-Q and D_1 pins together). The input clock pulses to this divider need rise times of less than 15 μS, and these are obtained via the push-button switches; each time a switch is closed C_1 charges rapidly, thus giving the fast-rise-time clock pulse. C_1 discharges slowly via R_2 when the switch reopens, eliminating false-triggering via switch-bounce effects, etc. Q_1 and the relay thus reliably change state each time a push-button switch is operated.

Circuits of the *Figure 2.18* type are of particular value in enabling hall, landing, or corridor lights to be controlled from several different switching points. In this case the push-button control switches can be connected to the main unit via very thin twin cable that can easily be hidden from sight. In this application the unit should (naturally) be powered from the AC mains (power) line,

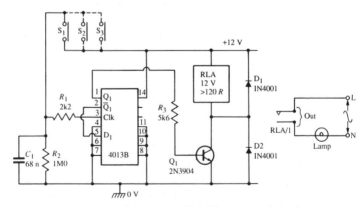

Figure 2.18   *Multi-input push-button on/off AC lamp control circuit*

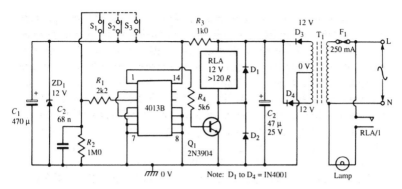

Figure 2.19 *Mains powered version of the push-button control circuit*

and *Figure 2.19* shows how it can be modified for this type of operation. Mains transformer T_1 must provide a 12 V-0-12 V output at 100 mA or greater.

Timer circuits

Relays can be used in many time-delayed-switching or timer applications, and *Figures 2.20* to *2.26* show some practical examples of circuits giving delays varying from a fraction of a second up to tens of hours.

Figures 2.20 to *2.22* show how inexpensive 4001B CMOS ICs can be used to implement medium-accuracy time delays of up to several minutes. The *Figure 2.20* design gives a delayed turn-on relay switching action, and operates as follows.

In *Figure 2.20* the CMOS gate is wired as a simple digital

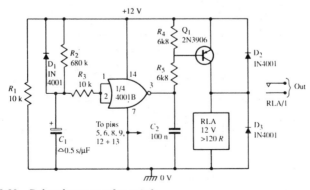

Figure 2.20 *Delayed-turn-on relay switch*

inverter, with its output driving the relay coil via pnp transistor Q_1 and with its input taken from the junction of time-controlled potential divider R_2–C_1. When power is first applied to the circuit C_1 is fully discharged, so the inverter input is grounded, its output is at full positive-rail potential, and Q_1 and the relay are off. As soon as power is applied C_1 starts to charge up via R_2 and applies a rising exponential voltage to the inverter input until, after a delay determined by the C_1–R_2 values, it reaches the inverter's threshold value, forcing its output to swing downwards and drive Q_1 and the relay on. The relay then remains on until circuit power is removed, at which point C_1 discharges rapidly via D_1 and R_1; the operating sequence is then complete.

Figure 2.21 shows how the above circuit action can be reversed,

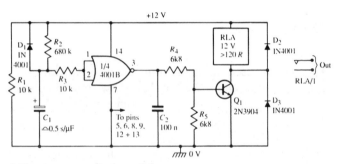

Figure 2.21   *Auto-turn-off relay switch*

so that the relay turns on as soon as power is applied but then turns off again after a pre-set delay, by simply modifying the relay-driving stage for npn transistor operation.

The above two circuits each give a delay of about 0.5 s/μF of C_1 value, enabling delays up to several minutes to be obtained; delays can be made variable by using a variable resistor in the R_2 position.

Figure 2.22 shows a pair of CMOS gates used to make a push-button-activated medium-accuracy one-shot relay switch that gives time delays up to several minutes. The action is such that the relay turns on as soon as start switch S_1 is briefly closed, but turns off again after a pre-set delay of about 0.5 s/μF of C_1 value. The two CMOS gates are simply wired as a manually-triggered monostable multivibrator that has its output fed to the relay via R_4 and Q_1.

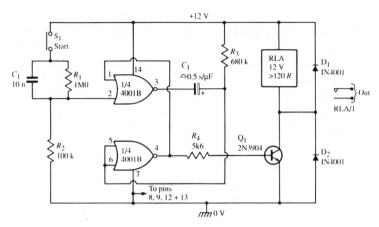

Figure 2.22 *One-shot relay time switch*

555 timer circuits

The *Figure 2.20* to *2.22* circuits are all based on simple CMOS elements, and are intended for use in medium-accuracy timer applications only. Far greater timing accuracy can be obtained by using a type-555 timer IC as the basic circuit timing element, and *Figures 2.23* to *2.26* show four practical high-accuracy timer circuits that are designed around this IC.

Figure 2.23 shows a simple 6-second to 60-second timer in which

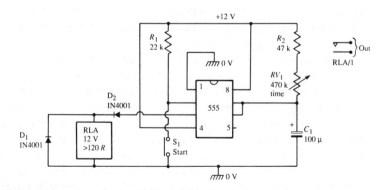

Figure 2.23 *Simple 6 to 60 seconds timer circuit*

the 555 is wired in the monostable or one-shot mode. A timing cycle is started by briefly closing S_1; relay RLA immediately turns on and C_1 starts to charge towards the positive rail via R_2 and RV_1 until eventually, after a delay determined by the RV_1 setting, C_1 rises to two thirds of the supply rail voltage, at which point the IC changes state and the relay turns off. The timing cycle is then complete.

Note that the above circuit draws supply current even when the relay is off. *Figure 2.24* shows a 2-range timer circuit that does not

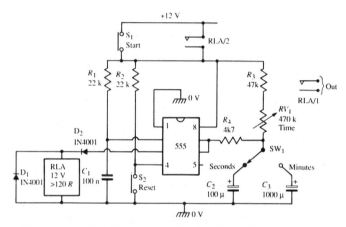

Figure 2.24   *2-range 6 to 60 seconds and 1 to 10 minutes timer*

suffer from this defect, and which covers the timing range 6-seconds to 10-minutes. The circuit operates as follows.

When S_1 is briefly closed a start pulse is fed to pin-2 of the IC via R_1 and C_1, and the relay turns on, contacts RLA/2 then change over and maintain the power connections when S_1 is released. The circuit then runs through a timing cycle like that described above, but with its timing period determined by either C_2 or C_3, until eventually the relay turns off and contacts RLA/2 reopen and break the circuit's supply connections. The timing cycle is then complete. Note that the circuit can be turned off part way through its timing cycle by operating reset switch S_2.

Precision timers

Conventional electrolytic capacitors have very wide tolerance values (typically −50 to +100 per cent), and have fairly large and unpredictable leakage currents. Consequently, simple circuits of the types shown in *Figures 2.23* and *2.24* can not be relied on to give precise timing periods or to give periods that exceed fifteen minutes or so. *Figures 2.25* and *2.26* show two high-accuracy long-period timers that do not rely on the use of the electrolytics for timing operations.

In these two circuits IC_1 is wired as a stable, variable frequency, free-running astable multivibrator. In the *Figure 2.25* design the

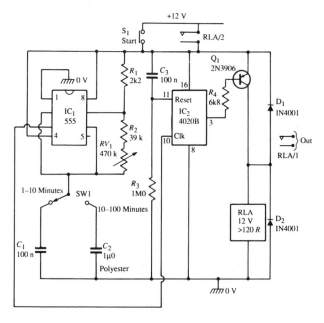

Figure 2.25   *2-range 1 to 10 minutes and 10 to 100 minutes timer*

astable frequency is divided down by IC_2, a 14-stage binary counter, so that the relay turns on as soon as S_1 is closed, and turns off again on the arrival of the 8192nd astable pulse, thereby giving total timing periods in the range 1 to 100 minutes.

The *Figure 2.26* circuit is similar, except that an additional decade-divider stage is used in position-3 of SW_1, thus giving a maximum division ratio of 81,920 and making timing periods of up to 20 hours available from the unit. This circuit is of particular value in giving time-controlled turn-off of battery chargers, etc.

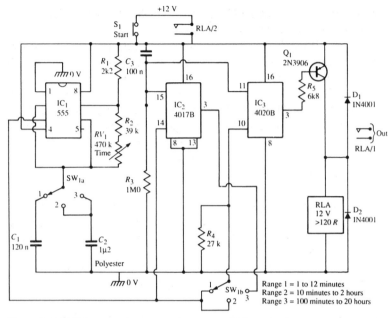

Figure 2.26   *Wide-range timer spans 1 minute to 20 hours in three ranges*

Relay pulsers

A relay pulser is a circuit that repeatedly turns a relay on and off for preset periods. *Figure 2.27* shows a simple version of such a circuit, wired as a lamp-flasher. Circuit operation relies on the fact that there is always a substantial difference between a relay coil's *pull-in* voltage (the value that just makes the contacts close) and its dropout voltage (at which the contacts reopen). Typically, a 12 V relay

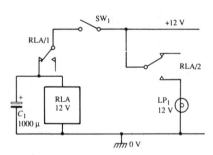

Figure 2.27   *Relay pulser or lamp flasher circuit*

may pull-in at about 10 V and drop-out at about 5 V. Thus, the *Figure 2.27* circuit operates as follows.

When SW_1 is first closed C_1 charges rapidly via closed contacts RLA/1 until the relay coil voltage reaches its pull-in value, at which point contacts RLA/1 open. C_1 then starts to discharge into the relay coil and thus holds the RLA/1 contacts open until the C_1 voltage falls to the relay's drop-out value, at which point contacts RLA/1 close again, and the whole timing process starts to repeat. Thus, the relay contacts repeatedly open and close at a rate determined by the C_1 and coil-resistance values and by the relative values of coil pull-in and drop-out voltages. The lamp (or other external circuitry) is activated via contacts RLA/2.

The simple relay pulser gives fairly poor timing accuracy, and calls for the use of a large value electrolytic timing capacitor. *Figure 2.28* shows a precision relay pulser that operates at a 1 Hz rate and uses a 10 μF timing capacitor. This design is based on a modified version of the standard '555 timer IC' astable multivibrator circuit.

In the standard form of this astable, C_1 first charges to two thirds of the supply-voltage value via R_3–R_4, at which point the 555 changes state and C_1 then discharges to one third of the supply-voltage value via R_4, at which point the 555 reverts to its original state and C_1 starts to recharge again via R_3–R_4, and so on ad infinitum. Thus, once the initial 'start-up' cycle is over, the C_1 voltage repeatedly swings between one third to two thirds of the supply value, and the IC generates a nearly-symmetrical square-wave output that switches the relay on and off at a 1 Hz rate. Note, however, that in the first 'start-up' half-cycle C_1 has to charge from

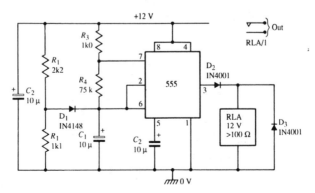

Figure 2.28 *'Precision' 1 Hz relay pulser*

zero to two thirds of the supply voltage, and this half-cycle is thus much longer than the succeeding ones. In *Figure 2.28* this fault is overcome via R_1–R_2 and D_1, which rapidly charge C_1 to one third of the supply voltage during the start-up half-cycle, thus giving a near-equal or 'precision' timing action on all operating half-cycles.

Photosensor basics

Relay switching circuits can easily be coupled to a photosensor and made to activate when light intensity goes above or below a preset value, or when a light beam is broken by a person or object, or when a light source is reflected onto the face of a light sensor by particles of smoke or fog, etc. The best known and easiest-to-use type of light-sensitive device (photosensor) is the LDR or 'light sensitive resistor', which uses the symbol shown in *Figure 2.29(a)*.

LDR operation relies on the fact that the resistance of a cadmium sulphide (CdS) film varies with the intensity of light falling on its face; the resistance is very high under dark conditions, and low under bright conditions. *Figure 2.29(b)* shows the LDR's basic construction, which consists of a pair of metal film contacts separated by a snake-like track of cadmium sulphide film designed to give the maximum possible contact area with the two metal films. The structure is housed in a clear plastic or resin case that gives free access to external light.

Practical LDRs are sensitive, inexpensive and readily available devices with good voltage and power handling capabilities, similar to those of a normal resistor. They are available in several sizes and package styles, the most popular size having a face diameter of roughly 10 mm. Typically, such a device has a resistance of several

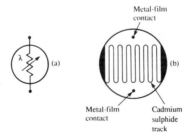

Figure 2.29   *(a) LDR symbol and (b) basic structure*

megohms under dark conditions, falling to about 900R at a light intensity of 100 Lux (typical of a well lit room) or about 30R at 8000 Lux (typical of bright sunlight).

Figures 2.30 to *2.36* show a selection of practical light-sensitive relay-output switching circuits that use LDRs as their optosensors; each of these circuits will work with virtually any LDR with a face diameter in the range 3 mm to 12 mm.

Light-activated relays

Figures 2.30 to *2.32* show a selection of relay-output light-activated 'switch' circuits based on the LDR. *Figure 2.30* shows a simple non-latching circuit that activates when light enters a normally-dark area such as the inside of a safe or cabinet, etc.

Here, R_1–LDR and R_2 form a light-sensitive potential divider that determines the base-bias of Q_1. Under dark conditions the LDR resistance is high, so zero base bias is fed to Q_1, and Q_1 and the relay are off. When a significant amount of light falls on the LDR face its resistance falls to a fairly low value, applying bias to the base of Q_1, which thus turns on and activates the RLA/1 relay contacts, which can be used to control external circuitry.

The simple *Figure 2.30* circuit has a fairly low sensitivity, and its 'trigger' points are fairly susceptible to changes in supply voltage and ambient temperature. *Figure 2.31* shows a CMOS-aided light-operated relay circuit that does not suffer from this susceptibility. Here, one of the four available NOR gates of a 4001B CMOS IC is

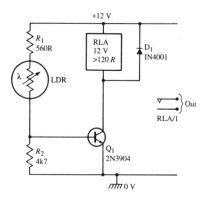

Figure 2.30   *Simple non-latching light-activated relay switch*

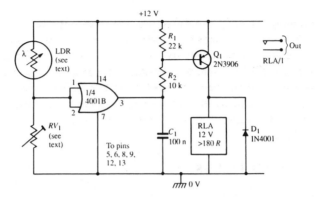

Figure 2.31   *CMOS-aided light-operated switch*

used as a linear inverter between the light-sensitive LDR-RV_1 network and the input of relay-driving transistor Q_1.

When the 4001B CMOS NOR gate is wired as an inverter it gives 'linear' action only when its input is within a few dozen millivolts of a value known as the 'threshold voltage' point; at all other times the gate output is saturated (driven to either ground or positive supply voltage). This 'threshold' point is actually a fixed fraction of the supply voltage value (the value is typically 50 per cent, but may vary from 30 per cent to 70 per cent between individual devices), and is very stable.

Thus, when it is wired in the configuration shown in *Figure 2.31*, the inverter input acts like the 'fixed potential divider' side of a Wheatstone bridge (with LDR–RV_1 forming the 'variable' side of the bridge) and acts as a bridge-balance detector that goes into the linear mode when the bridge is very close to its 'balance' point, which is not greatly influenced by variations in supply voltage or ambient temperature.

The action of this circuit is such that the gate input is low when the light level is below the desired 'trip' value, thus driving the gate output high and turning Q_1 and RLA off, but when the light level is above the trip value the gate input is high and its output is driven low, thus driving Q_1 and RLA on. When the light level is very close to the desired trip level the gate is driven into the linear mode, and small changes in light level can make the relay switch on or off. The trip level can be pre-set via RV_1.

The action of the above circuit can be reversed, so that the relay goes on when light intensity falls below a pre-set level, by transposing the RV_1 and LDR positions as in *Figure 2.32*, which also shows how a simple transient suppressing network can be wired between the output of the light-sensitive divider and the input of the CMOS gate, so that the circuit responds to mean light levels but is unaffected by sudden light transients (such as are caused by lightening flashes, etc.). This circuit can be used to automatically turn porch lights or car parking lights on at dusk and off at dawn.

Note in the above two circuits that the LDR can be any type that gives a resistance in the range 2k0 to 2M0 at the desired 'trip' level, and that (when adjusted) the RV_1 value should balance that of the LDR. Also note that C_1 is used to ensure stability of the CMOS inverter when it is operated in the linear mode.

Precision circuits

The CMOS-aided circuits of *Figure 2.31* and *2.32* give a semi-precision light-sensitive switching action that is adequate for most practical purposes. If even greater precision is needed it can be obtained by using the op-amp circuits of *Figures 2.33* to *2.35*.

In *Figure 2.33*, LDR–RV_1 and R_1–R_2 form a light-sensitive Wheatstone bridge that has its output fed to a sensitive 741 op-amp 'balance' detector. R_1–R_2 feed a fixed half-supply voltage to the detector's non-inverting input, and LDR–RV_1 feed a light-dependent voltage to its inverting input. If these two voltages differ by more than a few millivolts the op-amp output is driven to saturation (to near-zero or near-positive-rail values), thus driving Q_1 and RLA either on or off. If the two voltages are within a few millivolts of each other, the relay state depends on the direction of bridge imbalance. The actual balance point can be preset via RV_1, and is independent of variations in supply voltage or temperature. Because of the very high gain of the op-amp, the *Figure 2.33* circuit has a far greater sensitivity that the CMOS-aided circuits described earlier.

The *Figure 2.33* circuit is configured so that the relay turns on when the light goes above a preset level. The action can be reversed, so that it acts as a 'dark-activated' switch, by either transposing the inverting and non-inverting input connections of the op-amp, or by transposing the LDR and RV_1, as shown in *Figure 2.34*, which also shows how a small amount of hysteresis can

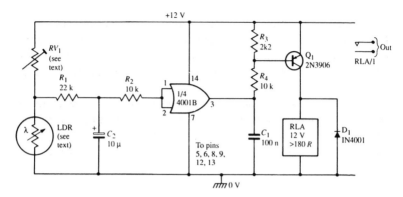

Figure 2.32 *CMOS-aided dark-operated switch with transient suppression*

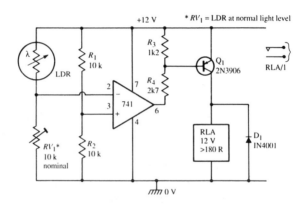

Figure 2.33 *Precision light-activated relay switch*

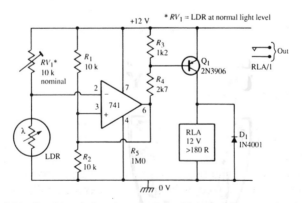

Figure 2.34 *Precision dark-activated switch with hysteresis*

be added via feedback resistor R_5, so that the relay turns on when the light level falls to a particular value but does not turn off again until it rises a substantial amount above this value. The hysteresis value is inversely proportional to the R_5 value, and is zero when R_5 is open circuit.

A precision combined light/dark switch, that activates a single relay if the light level goes above one preset value or below another, can be made by combining op-amp 'light' and 'dark' switches in the manner shown in *Figure 2.35*. To set up this circuit,

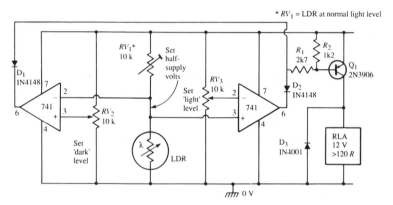

Figure 2.35 *Combined light-/dark-activated switch with single relay output*

first adjust RV_1 so that roughly half-supply volts appear on the RV_1–LDR junction when LDR is illuminated at the 'normal' or mean level. RV_2 can then be set so that RLA turns on when the light falls below the desired 'dark' level, and RV_3 can be set so that RLA turns on when the light rises above the desired 'light' or brightness level.

'Light-beam' alarms

One popular application of the LDR is as a light-beam alarm or switch, which activates when the passage of a beam of light is interrupted by an object or person. *Figure 2.36* shows a simple circuit of this type. The two lenses focus the lamp-generated light beam onto the LDR face, and the LDR circuit acts like a 'dark-operated' relay switch.

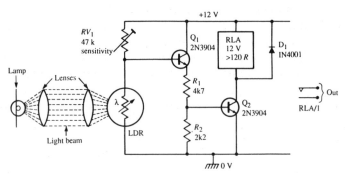

Figure 2.36   *Simple light-beam alarm with relay output*

Normally, with the light beam uninterrupted, the LDR face is illuminated by the beam and presents a low resistance, so little voltage appears at the RV_1-LDR junction and Q_2 and the relay are off. When the light beam is broken the LDR resistance increases, so a significant voltage appears on the RV_1–LDR junction and activates the relay via Q_2. In practice, most modern 'beam' alarms use a modulated infra-red signal to generate the 'light beam', and use an infra-red photodiode or transistor to detect the beam.

'Smoke' alarms

Another popular use of the LDR is as a 'smoke' alarm, which activates when smoke causes a light source to reflect onto the LDR face. *Figure 2.37* shows a sectional view of a reflective-type smoke

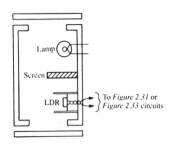

Figure 2.37   *Sectional view of reflective-type smoke detector*

detector, which consists of a lamp and an LDR mounted in an open ended but light-excluding box with an internal screen that stops the lamplight falling directly onto the LDR face. The lamp is a source of both light and heat, and the heat causes convection currents of air to be drawn in from the bottom of the box and expelled through the top. The inside of the box is painted matt black, and the construction lets air pass through ‍he box but excludes external light.

Thus, if the convected air currents are smoke free no light falls on the LDR face, and the LDR presents a high resistance, but if they do contain smoke the smoke particles make the lamp light reflect onto the LDR face and so cause a great and easily detectable decrease in the LDR resistance. The detector circuitry thus needs to act like a 'light-operated' circuit, and can take the form shown in *Figure 2.31* or *2.33*. Reflective smoke detectors of this type are far less susceptible to false alarms that conventional 'ionization' types of smoke alarms.

Heat-sensitive circuits

Each of the light-sensitive switching circuits of *Figures 2.30* to *2.36* can be converted into a temperature-sensitive switch by replacing its LDR with a negative-temperature-coefficient (NTC) thermistor. These are simple devices that present a resistance value inversely proportional to temperature, i.e., resistance falls as temperature rises, and vice versa.

Thus, *Figure 2.38* shows how the precision light-activated relay

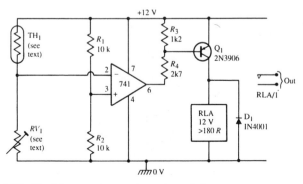

Figure 2.38 *Precision over-temperature relay switch*

switch of *Figure 2.33* can be converted into a precision over-temperature switch. The thermistor used here can be any NTC type that presents a resistance in the range 1k0 to 20k at the required trigger temperature, and the RV_1 resistance should equal this value at the same temperature. The above circuit can be made to act as an 'ice' or under-temperature switch by simply transposing TH_1 and RV_1.

Ordinary silicon diodes can be used as thermal sensors; they generate a volt drop of about 600 mV at a forward current of 1 mA and if this current is held constant the volt drop changes by about -2mV for each degree centigrade increase in diode temperature; all silicon diodes have inherently similar thermal characteristics. *Figure 2.39* shows how such a diode can be used as a sensor in an over-temperature switch.

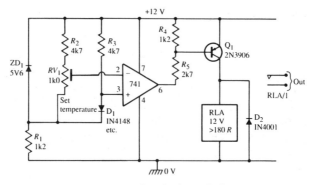

Figure 2.39   *Over-temperature switch with silicon diode sensor*

Here, ZD_1–R_1 generate a constant 5.6 V across the R_2–RV_1 and R_3–D_1 potential dividers, causing a virtual constant current to flow in each divider. A constant reference voltage is thus developed between the ZD_1–R_1 junction and pin-2 of the op-amp, and a temperature-dependent voltage with a coefficient of -2 mV/°C is developed between the ZD_1–R_1 junction and pin-3 of the op-amp. Thus, a differential NTC voltage appears between pins 2 and 3 of the op-amp.

To set up this circuit, simply apply the desired 'trip' temperature to D_1 and set RV_1 so that RLA just turns on. The circuit has an 'on/off' sensitivity of about 0.5 °C, and can be used as an overvalue switch at tempertures ranging from subzero to above the boiling point of water. Note that the circuit operation can be reversed, so that it acts as an under-temperature switch, by simply transposing the pin-2 and pin-3 connections of the op-amp.

Finally, *Figure 2.40* shows two silicon diodes used as thermal

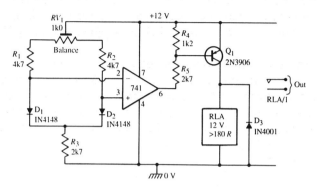

Figure 2.40 *Differential temperature switch*

sensors in a differential-temperature switch that turns on only when the D_2 temperature is more than a preset amount greater than that of D_1, and is not influenced by the absolute temperatures of the two diodes. The magnitude of this differential trip is fully variable from zero to about 10 °C via RV_1, so the circuit is quite versatile. It is set up by simply raising the D_2 temperature the required amount above that of D_1 and then carefully adjusting RV_1 so that the relay just turns on.

Miscellaneous circuits

To complete this chapter, *Figures 2.41* to *2.43* show three miscellaneous types of relay-switching circuits. The *Figure 2.41* design is that of another simple relay pulser, which repeatedly switches the relay on and off at a rate variable (via RV_1) between 26 and 80 cycles per minute via the CMOS astable multivibrator circuit built around the two 4001B NOR gates.

The above circuit can be used as an emergency lamp flasher by using its relay contacts to switch power to the lamps. *Figure 2.42* shows how it can be modified so that it starts pulsing or 'flashing' automatically when the ambient light level falls below a level preset via RV_2. Note that if this circuit is used as a lamp flasher care must

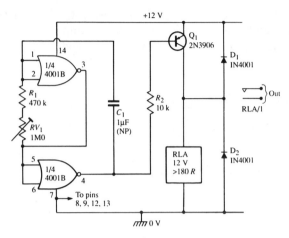

Figure 2.41 *Relay pulser circuit*

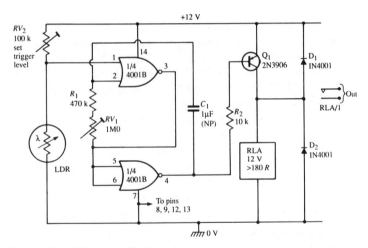

Figure 2.42 *Light-activated relay pulser*

be taken to ensure that the LDR face points away from the lamps, so that a photocoupled feedback loop is not set up.

Finally, *Figure 2.43* shows a water-activated relay switch that can be used to sound an alarm (via the relay contacts) when water reaches a pre-set level in a tank, bath, or other container. When the two metal probes are placed in water (or any other conductive liquid) the potential divider action of R_1 and the liquid resistance

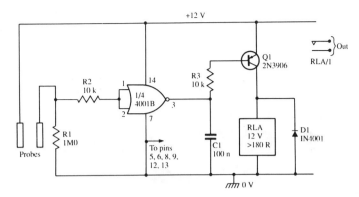

Figure 2.43 *Water-activated relay switch*

applies a 'high' voltage to the input of the inverter-connected CMOS NOR gate, driving its output low and activating RLA via Q_1. When the probes are not in a conductive liquid the gate's input is grounded via R_1, and the relay is off.

3 CMOS switches/ selectors

In Chapter 1 we took a brief look at CMOS bilateral switches and multiplexer-demultiplexer ICs, and explained that these devices can be used to replace conventional switches and relays in many low-level power control applications. In this chapter we expand on this theme and look at a variety of practical circuits.

CMOS bilateral switches

A CMOS bilateral switch or 'transmission gate' can be regarded as a near-perfect single-pole single-throw (SPST) electronic switch which, unlike an ordinary transistor or FET switch (which can conduct current in one direction only), can conduct analogue or digital signal currents in either direction (bilaterally) and can be turned on (closed) or off (opened) by applying a logic-1 or logic-0 signal to a single control terminal. Practical versions of such switches have a near-infinite off impedance, and a typical on impedance in the range 90 to 300 Ω.

Standard CMOS bilateral switches can be switched at frequencies ranging from zero to several megahertz, and have many uses in low-level power control circuits. They can, for example, be used to replace conventional mechanical switches and relays in signal-carrying applications, the bilateral switch being DC-controlled and placed directly on the printed circuit board (PCB) where it is needed, thus eliminating the problems of signal radiation and interaction that normally occur when such signals are mechanically switched via lengthy cables.

At high frequencies, CMOS bilateral switches can be used in

58

such diverse applications as signal gating, multiplexing, A–D and D–A conversion, digital control of frequency, impedance and signal gain, the synthesizing of multigang pots and capacitors, and the implementation of sample-and-hold circuits. Examples of most of these applications are shown later in this chapter.

Several types of CMOS 'multiple bilateral switch' ICs are available. These range from the simple 4016B and 4066B types, which each house four independently accessible SPST bilateral switches, to the 4097B, which houses an array of bilateral switches and logic networks arranged in the form of two independently-accessible single-pole 8-way bilateral switches or multiplexers/ demultiplexers. Before we take a closer look at the range of ICs, however, let us examine the basic operating principles and terminology, etc., of the CMOS bilateral switch.

Basic operating principles

Figure 3.1(a) show the basic circuit and *Figure 3.1(b)* the equivalent circuit of a simple CMOS bilateral switch. Here, an n-channel and p-channel MOSFET are wired in inverse parallel (drain-to-source and source-to-drain), but have their gate applied in antiphase from the control via a pair of CMOS inverter stages, to give the 'bilateral' switching action. Thus, when the control signal is at the logic-0 level the gate of Q_2 is driven to logic-1 and the gate of Q_1 is driven to logic-0, and under this condition both MOSFETs act as open switches between the X and Y points of the circuit. When, on the other hand, the control signal is set to the logic-1 level, the gate of

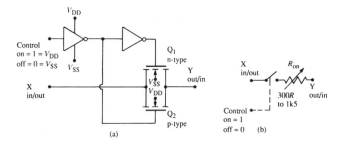

Figure 3.1 *(a) Basic circuit and (b) equivalent circuit of the simple CMOS bilateral switch*

Q_2 is driven to logic-0 and the gate of Q_1 is driven to logic-1, and under this condition both MOSFETs act as closed switches, and a low resistance (equal to the R_{SAT} value) exists between the circuit's X and Y points.

Note that, when the control input is at logic-1, signal currents can flow in either direction between the X and Y terminals (via Q_1 in one direction and Q_2 in the other), provided that the signal voltages are within the V_{SS}-to-V_{DD} power supply voltage values. Each of the X and Y terminals can thus be used as either an in or out signal terminal.

In practice, Q_1 and Q_2 have a finite resistance (R_{SAT}) when driven on, and in this simple circuit the R_{SAT} value may vary from 300R to 1k5, depending on the magnitude of the circuit's supply voltage and on the size and polarity of the actual input signal. This simple bilateral switch can thus be represented by the equivalent circuit of *Figure 3.1(b)*.

Figure 3.2 shows an improved version of the CMOS bilateral switch, together with its equivalent circuit. This is similar to the above, except for the addition of a second bilateral switch (Q_3–Q_4) that is wired in series with Q_5, with the 'well' of Q_1 tied to Q_5 drain. These modifications cause Q_1's well to switch to V_{SS} when the Q_1–Q_2 bilateral switch is off, but to be tied to the X input terminal when the switch is on. This modification reduces the on resistance of the Q_1–Q_2 bilateral switch to about 90R and virtually eliminates variations in its value with varying signal voltages, etc., as indicated by the equivalent circuit of *Figure 3.2(b)*. The only disadvantage of

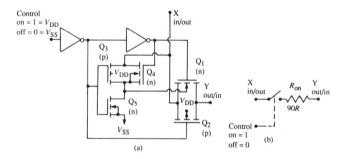

Figure 3.2 *(a) Basic circuit and (b) equivalent circuit of the improved CMOS bilateral switch*

the *Figure 3.2* circuit is that it has a slightly lower leakage resistance than *Figure 3.1.*

Switch biasing

A CMOS bilateral switch can be used to switch or gate either digital or analogue signals, but must be correctly biased to suit the type of signal being controlled. *Figure 3.3* shows the basic ways of activating and biasing the bilateral switch. *Figure 3.3(a)* shows that the switch can be turned on (closed) by taking the control terminal to V_{DD}, or turned off (open) by taking the control terminal to V_{SS}.

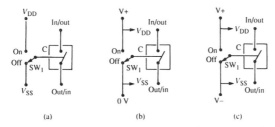

Figure 3.3 *(a) Basic methods of turning the bilateral switch on and off, and power supply connections for use with (b) digital and (c) analogue in/out signals*

In digital signal switching applications (*Figure 3.3(b)*) the bilateral switch can be used with a single ended supply, with V_{SS} to zero volts and V_{DD} at a positive value equal to (or greater than) that of the digital signal (up to a maximum of +18 V). In analogue switching applications (*Figure 3.3(c)*) a split supply (either true or effective) must be used, so that the signal is held at a mean value of 'zero' volts. The positive supply rail goes to V_{DD}, which must be greater than the peak positive voltage value of the input signal, and the negative rail goes to V_{SS} and must be greater than the peak negative value of the input signal. The supply values are limited to plus or minus 9 V maximum. Typically, the bilateral switch introduces less than 0.5 per cent of signal distortion when used in the analogue mode.

Logic-level conversion

Note from the above description of the analogue system that, if a split supply is used, the switch control signal must switch to the positive rail to turn the bilateral switch on, and to the negative rail to turn the switch off. This arrangement is inconvenient in many practical applications. Consequently, some CMOS bilateral switch ICs (notably the 4051B to 4053B family) have built-in 'logic-level conversion' circuitry which enables the bilateral switches to be controlled by a digital signal that switches between zero (V_{SS}) and positive (V_{DD}) volts, while still using split supplies to give correct biasing for analogue operation, as shown in *Figure 3.4*.

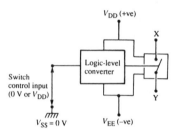

Figure 3.4   *Some ICs feature internal 'logic level conversion', enabling an analogue switch to be controlled via a single-ended input*

Multiplexing/demultiplexing

A multiplexer can be regarded as any system that enables information from a single *data* line to be distributed – on a sequential time/share basis – to a number (n) of independent data lines. *Figure 3.5*, for example, shows how a 4-way multiplexer (represented by a

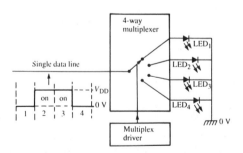

Figure 3.5   *A 4-way multiplexer used to control four LEDs via a single DATA line*

4-way switch) can be used to control (turn on or off) four LEDs down a single DATA line.

In *Figure 3.5*, assume that the multiplex driver continuously sequences the multiplexer through the '1–2–3–4' cycle at a fairly rapid rate, and is synchronized to the 1–2–3–4 segments of the data line. Thus, in each cycle, in the 1 period LED–1 is off; in the 2 period LED–2 is on; in the 3 period LED–3 is on, and in the 4 period LED–4 is off. The state of each of the four LEDs is thus controlled via the logic 'bit' of the single (sequentially time/shared) data line.

A demultiplexer is the opposite of a multiplexer. It enables information from a number (*n*) of independent data lines to be sequentially applied to a single data line. *Figure 3.6* shows how a 4-way demultiplexer can be used to feed three independent 'voice' signals down a single cable, and how a multiplexer can be used to convert signals back into three independent voice signals at the other end. In practice, each 'sample' period of the data line must be short relative to the period of the highest voice frequency; period-1 is used to synchronize the signals at the two ends of the data line.

From the above description it can be seen that a CMOS *n*-channel multiplexer can be regarded as a single-pole *n*-way bilateral switch, and that a CMOS multiplexer can be converted into a demultiplexer by simply transposing the notations of the input and output terminals. An *n*-way single-pole bilateral switch can thus be described as an *n*-channel multiplexer/demultiplexer.

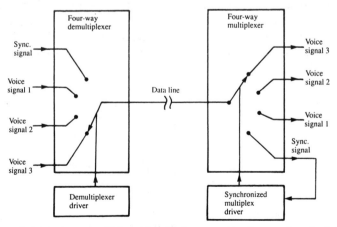

Figure 3.6 *A 4-way multiplexer/demultiplexer combination used to feed three independent voice signals through a single DATA line*

Practical ICs

There are three major families of CMOS bilateral-switch ICs. The best known of these are the 4016B/4066B types, which are quad bilateral switches, each housing four independently-accessible SPST bilateral switches, as already shown in Chapter 1 (in *Figure 1.27*). The 4016B uses the simple construction shown in *Figure 3.1*, and is recommended for use in sample-and-hold applications where low leakage impedance is of prime importance. The 4066B uses the improved type of construction of *Figure 3.2*, and is recommended for use in all applications where a low on resistance is of prime importance.

The second family of ICs are the 4051B to 4053B types (*Figures 1.28, 3.7*, and *3.8*). These are multichannel multiplexer/demultiplexer ICs featuring built-in logic-level conversion. These ICs have three 'power supply' terminals (V_{DD}, V_{SS}, and V_{EE}). In all applications, V_{DD} is taken to the positive supply rail, V_{SS} is grounded, and all digital control signals (for channel-select, inhibit, etc.) use these two terminals as their logic-reference values, i.e., logic-1 = V_{DD}, and logic-0 = V_{SS}. In digital-signal processing applications, terminal V_{EE} is grounded (tied to V_{SS}). In analogue-signal processing applications, V_{EE} must be taken to a negative supply rail. Ideally, $V_{EE} = -(V_{DD})$. In all cases, the V_{EE}-to-V_{DD} voltage must be limited to 18 V maximum.

The 4051B (*Figure 1.28*) is an 8-channel multiplexer/demultiplexer, and can be regarded as a single-pole, 8-way bilateral switch. The IC has three binary control inputs (A, B, and C) and an *inhibit* input. The three binary signals select the one of the eight channels to be turned on, as shown in the table.

The 4052B (*Figure 3.7*) is a differential 4-channel multiplexer/demultiplexer, and can be regarded as a ganged 2-pole, 4-way bilateral switch. It has two binary control inputs, which select the one of the four pairs of channels to be turned on, as shown in the table.

The 4053B (*Figure 3.8*) is a triple 2-channel multiplexer/demultiplexer, and can be regarded as a set of three independently-accessible single-pole 2-way bilateral switches, each controlled via a single terminal (A, B, or C).

The final family of devices are the 4067B and 4097B multiplexer/demultiplexer types (*Figures 3.9* and *3.10*). These devices can be used in both analogue and digital applications, but do not feature

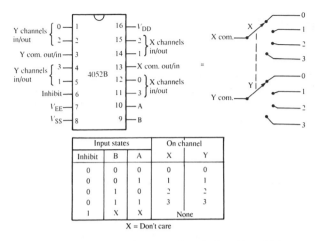

Figure 3.7 *The 4052B differential 4-channel multiplexer/demultiplexer acts as a ganged 2-pole 4-way switch*

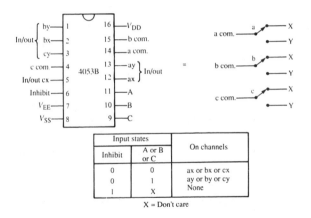

Figure 3.8 *The 4053B triple 2-channel multiplexer/demultiplexer acts as three independent single-pole 2-way switches*

built-in logic-level conversion. The 4067B is a 16-channel device, and can be regarded as a single-pole 16-way bilateral swtich. The 4097B is a differential 8-channel device, and can be regarded as a ganged 2-pole 8-way bilateral switch. Each IC is housed in a 24-pin DIL package.

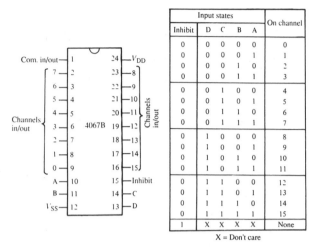

Figure 3.9   *The 4067B 16-channel multiplexer/demultiplexer acts as a single-pole 16-way switch*

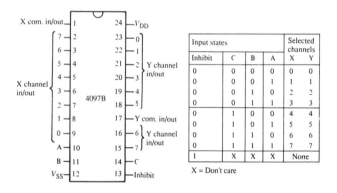

Figure 3.10   *The 4097B differential 8-channel multiplexer/demultiplexer acts as a ganged 2-pole 8-way switch*

Using 4016B/4066B ICs

The 4016B and 4066B are very versatile ICs, but the following simple precautions must be taken when using them:

1   Input and switching signals must never be allowed to rise above V_{DD} or below V_{SS}.
2   Each unused section of the IC must be disabled (see *Figure 3.11*) either by taking its control terminal to V_{DD} and wiring one

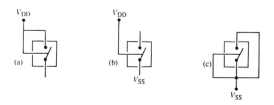

Figure 3.11 *Unused sections of the 4066B must be disabled, using any one of the connections shown here*

of its switch terminals to V_{DD} or V_{SS}, or by taking all three terminals to V_{SS}.

Figures 3.12 to *3.17* show some simple applications of the 4066B (or 4016B). *Figure 3.12* shows the device used to implement the

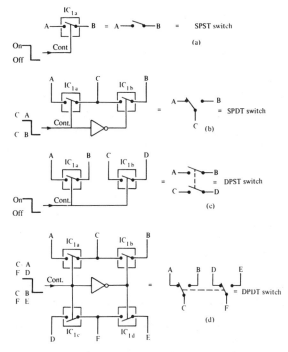

Figure 3.12 *Implementation of the four basic switching functions via the 4066B (IC₁)*

four basic switching functions of SPST, SPDT, DPST and DPDT. *Figure 3.12(a)* shows the SPST connection, which we have already discussed. The SPDT function (*Figure 3.12(b)*) is implemented by wiring an inverter stage (a 4001B or 4011B, etc.) between the IC_{1a} and IC_{1b} control terminals. The DPST switch (*Figure 3.12(c)*) is simply two SPST switches sharing a common control terminal, and the DPDT switch (*Figure 3.12(d)*) is two SPDT switches sharing the inverter stage in the control line.

Note that the switching functions of *Figure 3.12* can be expanded or combined in any desired way by using more IC stages. Thus, a 10-pole 2-way switch can be made by using five of the *Figure 3.12(d)* circuits with their control lines tied together.

Each 4066B bilateral switch has a typical on resistance of about 90*R*. *Figure 3.13* shows how four standard switch elements can be

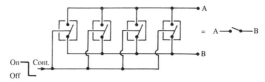

Figure 3.13   *This SPST switch has a typical on resistance of only 22.5 Ω*

wired in parallel to make a single switch with a typical on resistance of only 22.5 Ω.

Figures 3.14 to *3.17* show ways of using a bilateral switch as a self-latching device. In these circuits the switch current flows to ground via R_3, and the control terminal is tied to the top of R_3 via R_2. Thus, in *Figure 3.14*, when PB_1 is briefly closed the control terminal is

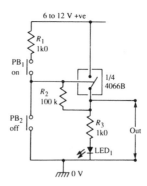

Figure 3.14   *Latching push-button switch*

pulled to the positive rail and the bilateral switch closes. With the bilateral switch closed, the top of R_3 is at supply-line potential and, since the control terminal is tied to R_3 via R_2, the bilateral switch is thus latched on. Once latched, the switch can only be turned off again by briefly closing PB$_2$, at which point the bilateral switch opens and the R_3 voltage falls to zero. Note here that LED$_1$ merely indicates the state of the bilateral switch, and R$_1$ prevents supply line shorts if PB$_1$ and PB$_2$ are both closed at the same time.

Figure 3.15 shows how the above circuit can be made to operate

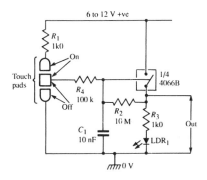

Figure 3.15 *Latching touch switch*

as a latching touch-operated switch by increasing R_2 to 10M and using R_4–C_1 as a 'hum' filter.

Figures 3.16 and *3.17* show different ways of using the *Figure 3.14* circuit to connect power to external circuitry. The *Figure 3.16*

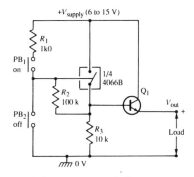

Figure 3.16 *Latching push-button power switch*

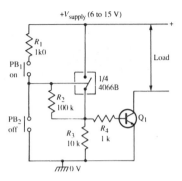

Figure 3.17   *Alternative version of the latching power switch*

circuit connects the power via a voltage follower stage, and the *Figure 3.17* design connects the power via a common-emitter amplifier.

Digital control circuits

Bilateral switches can be used to digitally control or vary effective values of resistance, capacitance, impedance, amplifier gain, oscillator frequency, etc., in any desired number of discrete steps. *Figure 3.18* shows how the four switches of a single 4066B can, by

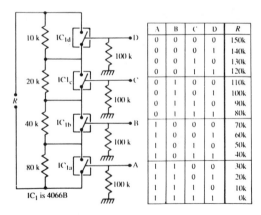

A	B	C	D	R
0	0	0	0	150k
0	0	0	1	140k
0	0	1	0	130k
0	0	1	1	120k
0	1	0	0	110k
0	1	0	1	100k
0	1	1	0	90k
0	1	1	1	80k
1	0	0	0	70k
1	0	0	1	60k
1	0	1	0	50k
1	0	1	1	40k
1	1	0	0	30k
1	1	0	1	20k
1	1	1	0	10k
1	1	1	1	0k

Figure 3.18   *16-step digital control of resistance. R can be varied from zero to 150k in 10k steps*

either shorting or not-shorting individual resistors in a chain, be used to vary the effective total value of the resistance chain in sixteen digitally-controlled steps of 10k each.

In practice, the step magnitudes of the *Figure 3.18* can be given any desired value (determined by the value of the smallest resistor) so long as the four resistors are kept in the ratio 1:2:4:8. The number of steps can be increased by adding more resistor/switch stages. Thus, a six-stage circuit (with resistors in the ratio 1:2:4:8:16:32) will give resistance variation in sixty-four steps.

Figure 3.19 shows how four switches can be used to make a digitally-controlled capacitor that can be varied in sixteen steps of 1n0 each. Again, the circuit can be expanded to give more 'steps' by simply adding more stages.

Note in the *Figure 3.18* and *3.19* circuits that the resistor/capacitor values can be controlled by operating the 4066B switches manually, or automatically, via simple logic networks, or via up/down counters, or via microprocessor control, etc.

The circuits of *Figure 3.18* and *3.19* can be combined in a variety of ways to make digitally-controlled impedance and filter networks, etc. *Figure 3.20*, for example, shows two alternative ways of using them to make a digitally-controlled first-order low-pass filter.

Digital control of amplifier gain can be obtained by hooking the *Figure 3.18* circuit into the feedback or input path of a standard op-amp inverter circuit, as shown in *Figures 3.21* and *3.22*. The gain of

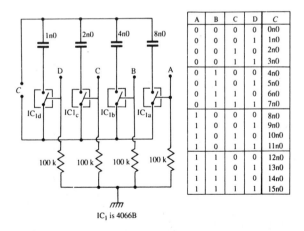

A	B	C	D	C
0	0	0	0	0n0
0	0	0	1	1n0
0	0	1	0	2n0
0	0	1	1	3n0
0	1	0	0	4n0
0	1	0	1	5n0
0	1	1	0	6n0
0	1	1	1	7n0
1	0	0	0	8n0
1	0	0	1	9n0
1	0	1	0	10n0
1	0	1	1	11n0
1	1	0	0	12n0
1	1	0	1	13n0
1	1	1	0	14n0
1	1	1	1	15n0

Figure 3.19 *16-step digital control of capacitance. C can be varied from zero to 15n in 1n0 steps*

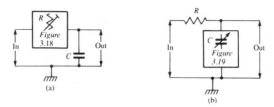

Figure 3.20   *Alternative ways of using Figure 3.18 or 3.19 to make a digitally-controlled first-order low-pass filter*

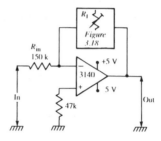

Figure 3.21   *Digital control of gain, using the Figure 3.18 circuit. Gain is variable from zero to unity in sixteen steps*

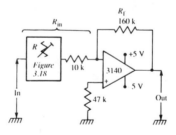

Figure 3.22   *Digital control of gain, using the Figure 3.18 circuit. Gain is variable from unity to 16 in sixteen steps*

such a circuit equals R_f/R_{in}, where R_f is the feedback resistance and R_{in} is the input resistance. Thus, in *Figure 3.21*, the gain can be varied from zero to unity in sixteen steps of 1/15th each, giving a sequence of 0/15 (i.e., zero), 1/15, 2/15, etc., up to 14/15 and (finally) 15/15 (i.e., unity).

   In the *Figure 3.22* circuit the gain can be varied from unity to 16 in sixteen steps, giving a gain sequence of 1, 2, 3, 4, 5, etc. Note in both of these circuits that the op-amp uses a split power supply, so the 4066B control voltage must switch between the negative and positive supply rails.

Figure 3.23 shows how the *Figure 3.18* circuit can be used to vary

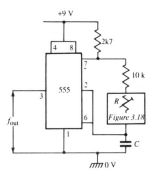

Figure 3.23   *Digital control of 555 astable frequency, in sixteen steps*

the frequency of a 555 astable oscillator in sixteen discrete steps. Finally, *Figure 3.24* shows how three bilateral switches can be used to implement digital control of decade range selection of a 555 astable oscillator. Here, only one of the switches must be turned on at a time. Naturally, the circuits of *Figure 3.23* and *3.24* can easily

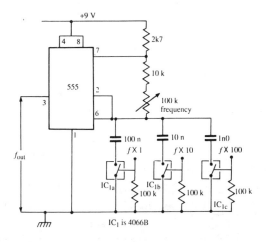

Figure 3.24   *Digital control of decade range selection of a 555 astable*

be combined, to form a wide-range oscillator that can be digitally controlled via a micro, etc.

Synthesized multigang pots, etc.

One of the most useful applications of the bilateral switch is in synthesizing multigang rheostats, potentiometers (pots), and variable capacitors in a.c. signal-processing circuitry. The synthesizing principle is quite simple and is illustrated in *Figure 3.25*, which shows the circuit of a 4-gang 10k-to-100k rheostat for use at signal frequencies up to about 15 kHz.

Here, the 555 is used to generate a 50 kHz rectangular waveform that has its mark/space (M/S) ratio variable from 11:1 to 1:11 via RV_1, and this waveform is used to control the switching of the 4066B stages. All of the 4066B switches are fed with the same control waveform, and each switch is wired in series with a range resistor (R_a, R_b, etc.), to form one gang of the rheostat between the a–a, b–b, etc., terminals.

Remembering that the switching rate of this circuit is fast (50 kHz) relative to the intended maximum signal frequency (15 kHz), it can be seen that the *mean* or effective value (when integrated over a few switching cycles) of each 'rheostat' resistance can be varied via M/S-ratio control RV_1.

Thus, if IC_{2a} is closed for 90 per cent and open for 10 per cent of each duty cycle (M/S-ratio = 9:1), the apparent (mean) value of the

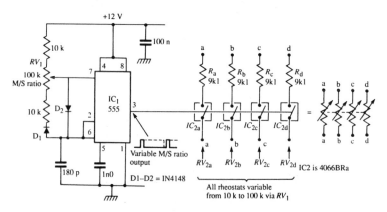

Figure 3.25   *Synthesized precision 4-gang 'rheostat'*

a–a resistance will be 10 per cent greater than R_a, i.e., 10k. If the duty cycle is reduced to 50 per cent, the apparent R_a value doubles to 18.2k. If the duty cycle is further decreased, so that IC_{2a} is closed for only 10 per cent of each duty cycle (M/S-ratio = 1:9), the apparent value of R_a increase by a decade to 91k. Thus, the apparent value of each gang of the 'rheostat' can be varied via RV_1.

There are some important points to note about the above circuit. First, it can be given any desired number of 'gangs' by simply adding an appropriate number of switch stages and range resistors. Since all switches are controlled by the same M/S-ratio waveform, perfect tracking is automatically assured between the gangs. Individual gangs can be given different ranges, without effecting the tracking, by giving them different range resistor values. Also, the 'sweep' range and 'law' of the rheostats can be changed by simply altering the characteristics of the M/S-ratio generator. Finally, note that the switching control frequency of the circuit *must be* far higher than the maximum signal frequency that is to be handled, or the circuit will not perform correctly.

The 'rheostat' circuit of *Figure 3.25* can be made to function as a multigang variable capacitor by using ranging capacitors in place of ranging resistors. In this case, however, the apparent capacitance value decreases as the duty cycle is decreased.

The *Figure 3.25* principle can be expanded to make synthesized multigang pots by using the basic technique shown in *Figure 3.26*. Here, in each gang, two 'rheostats' are wired in series but have their switch control signals fed in antiphase, so that one rheostat value increases as the other decreases, thus giving a variable potential-

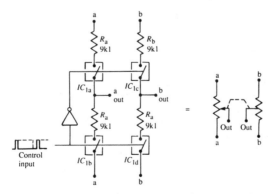

Figure 3.26 *Synthesized precision 2-gang 'potentiometer'*

divider action. This basic circuit can be expanded to incorporate any desired number of gangs, by simply adding more double-rheostat stages.

Miscellaneous applications

To complete this look at the CMOS bilateral switch, *Figure 3.27* and *3.28* show a couple of miscellaneous fast-switching applications of the device. *Figure 3.27* shows how it can be used as a sample-and-

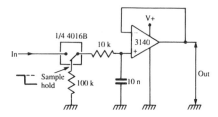

Figure 3.27   *Using a bilateral switch as a sample-and-hold element*

hold element. The 3140 CMOS op-amp is used as a voltage follower and has a near-infinite input impedance, and the 4016B switch also has a near-infinite impedence when open. Thus, when the 4016B is closed, the 10n capacitor rapidly follows all variations in input voltage, but when the switch opens the prevailing capacitance-charge is stored and the resulting voltage remains available at the op-amp output.

Finally, *Figure 3.28* shows a bilateral switch used in a linear ramp generator. Here, the op-amp is used as an integrator, with its non-

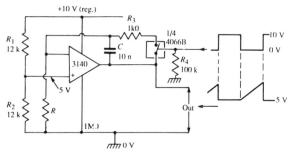

Figure 3.28   *Using the 4066B to implement a ramp generator circuit*

inverting pin biased at 5 V via R_1–R_2, so that a constant current of 5 μA flows into the inverting pin via R. When the bilateral switch is open, this current linearly charges capacitor C, causing a rising ramp to appear at the op-amp output. When the bilateral switch closes, C is rapidly discharged via R_3 and the output switches down to 5 V.

The basic *Figure 3.28* circuit is quite versatile. The switching can be activated automatically via a free-running oscillator or via a voltage-trigger circuit. Bias levels can be shifted by changing the R_1–R_2 values or can be switched automatically via another 4066B stage, enabling a variety of ramp waveforms to be generated.

4 AC power control circuits

The easiest way of electronically controlling the power feed to lamps, heaters, motors, or other devices that are operated from the standard 50 Hz to 60 Hz AC power line is via solid-state thyristor devices such as SCRs or triacs. The basic operating theory and application principles of these devices have already been extensively described in Chapter 1, and in this chapter we concentrate on practical AC power control applications of these devices.

A total of twenty-eight practical circuits are shown in this chapter, and each one can be used on either 115 V or 230 V AC power lines (where applicable, 115 V component values are shown in parentheses). In each design, users must simply select the triac or SCR rating to suit their own particular application. Let us start off, then, by looking at some simple triac circuits that can be used to switch AC power to lamps, heaters, motors, or any of a variety of domestic or industrial appliances.

Triac power switches

Triacs are, as already explained in Chapter 1, solid state power switches that can be triggered (turned on and latched) either synchronously or non-synchronously with the AC mains voltage, but which automatically turn off at the end of each mains half-cycle as their main-terminal currents fall below the device's 'minimum holding' value.

When used as simple power switches, synchronous circuits *always* turn on at the same point in each AC half-cycle (usually just after the zero-crossing point) and can generate minimal RFI. The

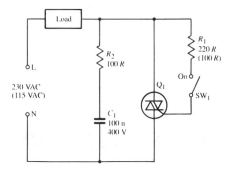

Figure 4.1 *Simple AC power switch, AC line triggered*

trigger points of non-synchronous circuits, on the other hand, are not invariably synchronized to a fixed point of the AC cycle, and may generate significant RFI, particularly at the point of initial turn-on.

Figures 4.1 to *4.8* show a variety of non-synchronous triac power switch circuits that can be used in basic on/off line switching applications. The basic action of the *Figure 4.1* circuit was explained in Chapter 1, and is such that the triac is off and acts like an open switch when SW_1 is open, but acts like a closed switch that is gated on from the mains via the load and R_1 shortly after the start of each AC half-cycle when SW_1 is closed. Note that the triac's main terminal voltage drops to only a few hundred millivolts as soon as the triac turns on, so R_1 and SW_1 consume very little mean power, and that the triac's trigger point is *not* synchronized to the mains when SW_1 is initially closed, but becomes synchronized on all subsequent half-cycles. Also note that R_2–C_1 form a 'snubber'

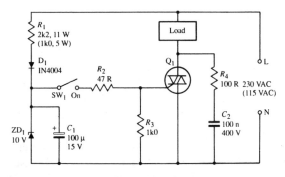

Figure 4.2 *AC power switch with line-derived DC triggering*

network that provides the triac with rate-effect suppression; similar networks are fitted to most triac circuits shown here.

Figure 4.2 shows how the triac can be used as a power switch that can be triggered via a mains-derived DC supply. C_1 is charged to +10 V on each positive mains half-cycle via R_1–D_1 and the C_1 charge triggers the triac when SW_1 is closed. Note that R_1 is subjected to almost the full AC line voltage at all times, and thus needs a fairly high power rating, and that all parts of this circuit are 'live', making it difficult to interface to external control circuitry.

Isolated-input control

Figure 4.3 shows how the above circuit can be modified so that it can easily be interfaced to external control circuitry. Here, SW_1 is simply replaced by transistor Q_2, which in turn is driven from the phototransistor side of an inexpensive optocoupler. The LED side of the optocoupler is driven from a 5 V (or greater) DC supply via R_4. The triac turns on only when the external supply is connected via SW_1.

Optocouplers have insulation potentials as high as several thousand volts, so the above external circuit is fully isolated from the mains-driven triac circuitry, and can easily be designed to give any desired form of automatic 'remote' triac operation by replacing SW_1 with a suitable electronic switch.

Figure 4.4 shows a variation of the above circuit. Here, the triac is AC triggered in each half-cycle via the AC impedance of C_1–R_1 and via back-to-back zeners ZD_1–ZD_2, but C_1 dissipates near-zero power. Bridge rectifier D_1-to-D_4 is wired across the ZD_1–ZD_2–R_2 network and is loaded by Q_2; when Q_2 is off, the bridge is

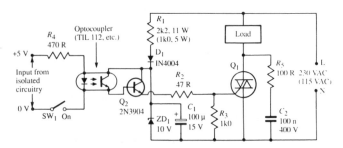

Figure 4.3   *Isolated-input (optocoupled) AC power switch, DC triggered*

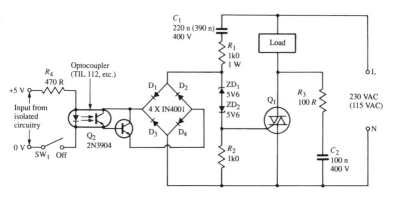

Figure 4.4 *Isolated-input AC power switch, AC triggered*

effectively open and the triac turns on just after the start of each AC half-cycle, but when Q_2 is on a near-short appears across ZD_1–ZD_2–R_2, so the triac is off. Q_2 is actually driven via the optocoupler from the isolated external circuit, so the triac is normally on but turns off when SW_1 is closed.

DC triggering

Figure 4.5 and *4.6* show ways of triggering a triac power switch via a transformer-derived DC supply and a transistor-aided switch. In *Figure 4.5* the transistor and the triac are both driven on when SW_1 is closed, and are off when SW_1 is open. In practice, SW_1 can easily be replaced by an electronic switch, enabling the triac to be operated by heat, light, sound, time, etc. Note, however, that the whole of this circuit is 'live'; *Figure 4.6* shows the circuit modified

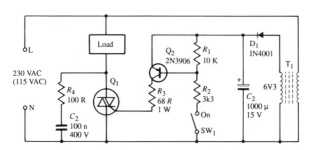

Figure 4.5 *AC power switch with transistor-aided DC triggering*

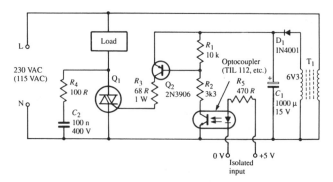

Figure 4.6   *Isolated-input AC power switch with DC triggering*

for optocoupler operation, enabling it to be activated via fully-isolated external circuitry.

UJT triggering

To complete this look at non-synchronous triac on/off power switching circuitry, *Figures 4.7* and *4.8* show two ways of obtaining triac triggering via a fully isolated external circuit. The triggering action is obtained via UJT (unijunction transistor) oscillator Q_2, which operates at several kilohertz and feeds output pulses to the triac gate via pulse transformer T_1, which provides the desired 'isolation'. Because of its fairly high oscillating frequency, the UJT triggers the triac within a few degrees of the start of each mains half-cycle when the oscillator is active.

   In *Figure 4.7*, Q_3 is wired in series with the UJT's main timing resistor, so the UJT and triac turn on only when SW_1 is closed. In

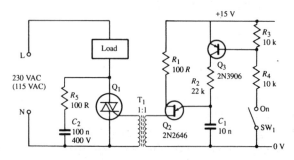

Figure 4.7   *Isolated-input (transformer coupled) AC power switch*

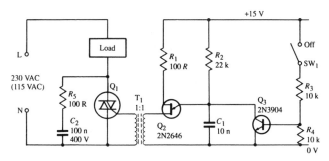

Figure 4.8 *Isolated-input AC power switch*

Figure 4.8, Q_3 is wired in parallel with the UJT's main timing capacitor, so the UJT and triac turn on only when SW_1 is open. In each case, SW_1 can easily be replaced by an electronic switch, to give some form of automatic power switching action.

Automatic control

The main advantage of the *Figure 4.3* to *4.8* triac circuits, when compared to ordinary electromechanical switching circuits, is that they can easily be modified to give automatic switching action in response to variations in time, light, or heat, etc., by simply using suitable circuitry in the input control position. An almost infinite variety of such control circuits can easily be devised, and many of the relay-output circuits of Chapter 2 can be modified to give direct triac activation.

Synchronous power switching

Synchronous AC power switches can easily be designed to generate minimal RFI, and *Figures 4.9* to *4.18* show a variety of synchronous triac power switch circuits that can be used in various on/off line switching applications.

Figure 4.9 shows a 'transistorized' synchronized AC power switch that is triggered near the zero-voltage cross-over points of the AC waveform. The triac gate trigger current is obtained from a 10 V DC supply derived from the mains via R_1–D_1–ZD_1 and C_1, and this supply is switched to the gate via Q_5, which in turn is

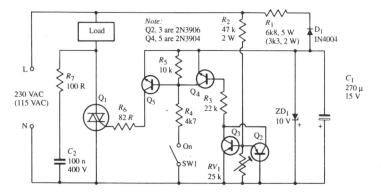

Figure 4.9 *'Transistorized' synchronous line switch*

controlled by SW_1 and zero-crossing detector Q_2–Q_3–Q_4. The action of Q_5 is such that it can turn on and conduct gate current when SW_1 is closed and Q_4 is off. The action of the zero-crossing detector is such that Q_2 or Q_3 are driven on whenever the instantaneous mains voltage is positive or negative by more than a few volts (depending on the setting of RV_1), thereby driving Q_4 on via R_3 and inhibiting Q_5. Thus, gate current can only be fed to the triac when SW_1 is closed and the instantaneous mains voltage is within a few volts of zero; this circuit thus generates minimal switching RFI.

Figure 4.10 shows how the above circuit can be modified so that the triac can only turn on when SW_1 is open. Note in both circuits that, since only a narrow pulse of gate current is sent to the triac, the *mean* consumption of the DC supply is very low (about 1 mA). Also note that SW_1 can easily be replaced by an electronic switch,

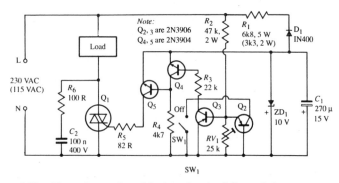

Figure 4.10 *Alternative version of the 'transistorized' line switch*

to give automatic operation via heat, light, time, etc., or by an optocoupler to give fully isolated operation from external circuitry.

Special ICs

Several dedicated synchronous zero-crossover triac-gating ICs are available. The best known of these are the CA3059 and the TDA1024, which each have built-in mains-derived DC power supply circuitry, a zero-crossing detector, triac gate drive circuitry, and a high-gain differential amplifier/gating network.

Figure 4.11 shows the internal circuitry of the CA3059, together with its minimal external connections. Mains power is connected to pins 5 and 7 via limiting resistor R_s (22 k, 5 W when 230 V mains is used). D_1 and D_2 act as back-to-back zeners and limit the pin 5

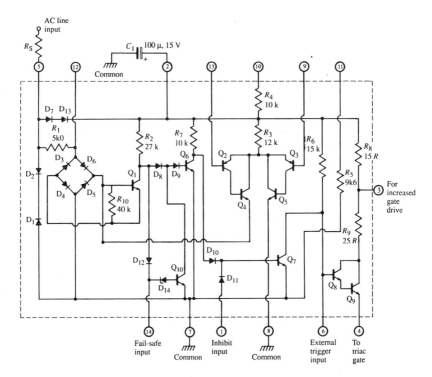

Figure 4.11 *Internal circuit and minimal external connections of the CA3059 synchronous 'zero-voltage' triac driver IC*

voltage to ±8 V. On positive half cycles D_7 and D_{13} rectify this voltage and generate 6.5 V across the 100 μF pin-2 capacitor, which stores enough energy to drive all internal circuitry and provides adequate Triac gate drive, with a few milliamperes of spare drive available for powering external circuitry if needed.

Bridge rectifier D_3 to D_6 and transistor Q_1 act as a zero-crossing detector, with Q_1 being driven to saturation whenever the pin 5 voltage exceeds ±3 V. Gate drive to an external triac can be made via the emitter (pin 4) of the Q_8–Q_9 Darlington pair, but is available only when Q_7 is turned off. When Q_1 is turned on (pin 5 greater than ±3 V) Q_6 turns off through lack of base drive, so Q_7 is driven to saturation via R_7 and no triac gate drive is available at pin 4. Triac gate drive is thus available only when pin 5 is close to the 'zero-voltage' mains value, and is delivered in the form of a narrow pulse centred on the cross over point; its power is supplied via C_1.

The CA3059 incorporates a general-purpose voltage comparator or differential amplifier, built around Q_2 to Q_5. Resistors R_4 and R_5 are externally available for biasing one side of the amplifier. The emitter current of Q_4 flows via the base of Q_1 and can be used to disable the triac gate drive (pin 4) by turning Q_1 on. The configuration is such that the gate drive can be disabled by making pin 9 positive relative to pin 13. The drive can also be disabled by connecting external signals to pin 1 and/or pin 14.

Figures 4.12 and *4.13* show how the CA3059 can be used to give manually controlled 'zero voltage' on/off switching of the triac. These two circuits use SW_1 to enable or disable the triac gate drive via the IC's internal differential amplifier (the drive is enabled only when pin 13 is biased above pin 9). In the *Figure 4.12* circuit pin 9 is

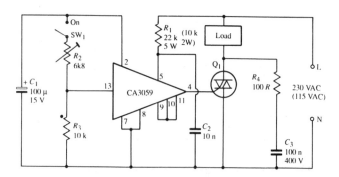

Figure 4.12 *Direct-switching IC-gated 'zero-voltage' line switch*

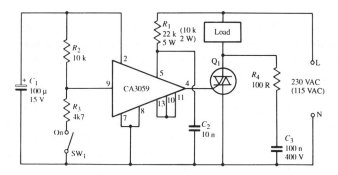

Figure 4.13 *An alternative method of direct-switching the CA3059 IC*

biased at half-supply volts and pin 13 is biased via R_2–R_3 and SW_1, and the triac thus turns on only when SW_1 is closed.

In *Figure 4.13*, pin 13 is biased at half-supply volts and pin 9 is biased via R_2–R_3 and SW_1, and the Triac again turns on only when SW_1 is closed. In both circuits, SW_1 handles maximum potentials of 6 V and maximum currents of about 1 mA. Note in these designs that capacitor C_2 is used to apply a slight phase delay to the pin 5 'zero voltage detecting' terminal, and causes the gate pulses to occur after (rather than to 'straddle') the zero-voltage point.

Note in the *Figure 4.13* circuit that the triac can be turned on by pulling R_3 low or can be turned off by letting R_3 float. *Figures 4.14* and *4.15* show how this simple fact can be put to use to extend the versatility of the basic circuit. In *Figure 4.14* the triac can be turned

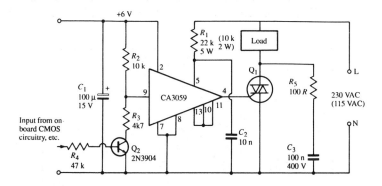

Figure 4.14 *Method of transistor-switching the CA3059 via on-board CMOS circuitry, etc.*

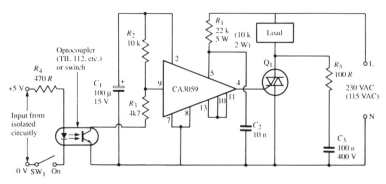

Figure 4.15   *Method of remote-switching the CA3059 via an optocoupler*

on and off by transistor Q_2, which in turn can be activated by on-board CMOS circuitry (such as one-shots, astables, etc.) that are powered from the 6 V pin 2 supply.

In *Figure 4.15* the circuit can be turned on and off by fully isolated external circuitry via an inexpensive optocoupler, which needs an input in excess of only a couple of volts to turn the triac on.

Alternatively, *Figure 4.16* shows how the TDA1024 IC can be used in place of the CA3059 to give either directly switched or optocoupled 'zero-voltage' triac control.

Finally, *Figure 4.17* and *4.18* show a couple of ways of using the CA3059 so that the triac operates as a light-sensitive 'dark-operated' power switch. In these designs the IC's built-in differential amplifier is used as a precision voltage comparator that turns the Triac on or off when one of the comparator input voltage goes above or below the other.

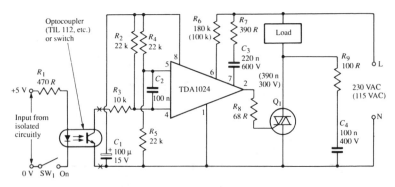

Figure 4.16   *The TDA1024 used to give either directly switched or optocoupled 'zero-voltage' triac control*

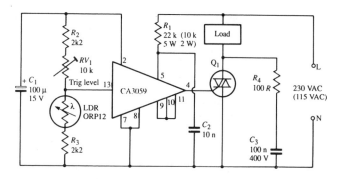

Figure 4.17   *Basic 'dark-activated' zero-voltage switch*

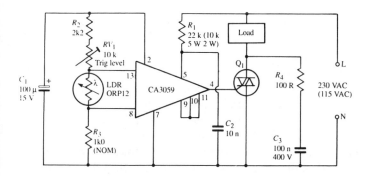

Figure 4.18   *Dark-activated zero-voltage switch with hysteresis provided via R_3*

Figure 4.17 shows a simple dark-activated power switch. Here, pin 9 is tied to half-supply volts and pin 13 is controlled via the R_2–RV_1–LDR–R_3 potential divider. Under bright conditions the LDR has a low resistance, so pin 13 is below pin 9 and the triac is disabled. Under dark conditions the LDR has a high resistance, so pin 13 is above pin 9 and the triac is enabled and power is fed to the AC load. The threshold switching level can be preset via RV_1.

Figure 4.18 shows how a degree of hysteresis or 'backlash' can be added to the above circuit, so that the triac does not switch annoyingly in response to small changes (such as caused by passing shadows, etc.) in ambient light level. The hysteresis level is controlled via R_3, which can be selected to suit particular applications.

Electric heater controllers

Triacs can be used to give automatic room temperature control by using electric heaters as the triac loads, and thermostats or thermistors as the triac feedback control elements. Two basic methods of heater control can be used, either simple automatic on/ off power switching, or fully automatic 'burst-fire' proportional power control, as described in Chapter 1. In the former case, the heater switches fully on when the room temperature falls below a preset level and turns off when it rises above the preset level. In the latter (burst-fire) case, the *mean* heater power is automatically adjusted so that, when the room temperature is at the precise pre-set level, the heater output power self-adjusts to exactly balance the thermal losses of the room.

Because of the high power requirements of electric heaters, special care must be taken in the design of triac controllers to keep RFI-generation to minimal levels. Two options are open to designers, to use continuous DC gating of the triac, or to use synchronously pulsed gating. The advantage of DC gating is that, in basic on/off switching applications, the triac generates zero RFI under normal (on) running conditions; the disadvantage is that the triac may generate a very powerful RFI spike as it is initially switched from the off to the on condition.

The advantage of synchronous gating is that no high-level RFI is generated as the triac transitions from the off to the on condition; the disadvantage is that the triac generates continuous very low level RFI under normal (on) running conditions.

DC-gated circuits

Figures 4.19 and *4.20* show examples of DC-gated heater controller circuits, in which the DC supply is derived via T_1–D_1 and C_1, and the heater can be controlled either manually or automatically via SW_1. The *Figure 4.19* circuit is autocontrolled via a thermostat, and calls for no further explanation.

The *Figure 4.20* circuit, on the other hand, is controlled by NTC thermistor TH_1 and transistors Q_2–Q_3, and calls for some explanation. RV_1–TH_1–R_2–R_3 are used as a thermal bridge, with Q_2 acting as a bridge-balance detector. RV_1 is adjusted so that Q_2 just starts to turn on as the temperature falls to the desired preset level; below

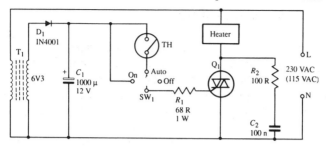

Figure 4.19 *Heater controller with thermostat-switched DC gating*

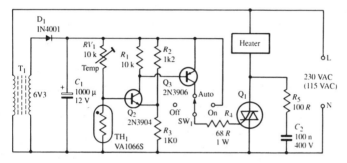

Figure 4.20 *Heater controller with thermistor-switched DC gating*

this level, Q_2–Q_3 and the triac are all driven hard on, and above this level all three components are cut off.

Note in this circuit that, since the gate-drive polarity is always positive but the triac main-terminal current is alternating, the triac is gated alternately in the so-called I+ and III+ modes or quadrants, and that the gate sensitivities are quite different in these two modes. Consequently, when the TH_1 temperature is well below the preset level Q_3 is driven hard on and the triac is gated in both quadrants and gives full power drive to the heater, but when the temperature is very close to the preset value Q_3 is only lightly driven on, so the triac is gated in the I+ mode only and the heater operates at only half of maximum power drive. The circuit thus gives fine control of temperature.

Synchronous circuits

Figure 4.21 shows how a CA3059 IC can be used to make an automatic thermistor-regulated synchronous electric heater con-

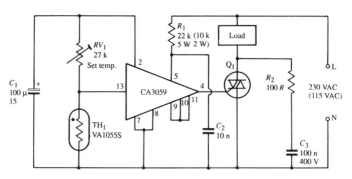

Figure 4.21 *Heater controller with thermistor-regulated zero-voltage switching*

troller that gives a simple on/off heater switching action. The circuit is similar to the 'dark-activated' power switch of *Figure 4.17* except the NTC thermistor TH_1 is used as the feedback sensing element. The circuit is capable of maintaining room temperature within a degree or so of the value preset via RV_1.

Finally, to complete this 'heater controller' section, *Figure 4.22* shows a 'burst-fire' or proportional heater controller that is capable of regulating room temperatures to within ±0.5 °C of a preset value. Here, a thermistor controlled voltage is applied to the pin-13 side of the CA3059's comparator, and a repetitive 300 mS ramp waveform, centred on half-supply volts, is applied to the pin-9 side of the comparator from CMOS astable ICI.

The circuit action is such that the triac is synchronously gated fully on if the ambient temperature is more than a couple of degrees below the preset level, or is cut off if it is more than a couple of degrees above the preset level, but when it is close to the preset value the ramp waveform comes into effect and synchronously

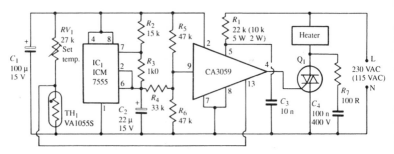

Figure 4.22 *Heater controller giving integral-cycle precision temperture regulation*

turns the triac on and off in the burst-fire or integral cycle mode once every 300 mS, with a mark/space (M/S)-ratio proportional to the temperature differential.

Thus, if the M/S-ratio is 1:1, the heater generates only half of maximum power, and if the ratio is 1:3 it generates only one quarter of maximum power. The net effect is that the heater output power self-adjusts to meet the room's heating needs. When the room temperature reaches the precise preset value, the heater does not switch fully off, but generates just enough output power to precisely match the thermal losses of the room. The system gives very precise temperature control.

AC lamp dimmer circuits

Triacs can easily be used to make very efficient lamp dimmers (which vary the brilliance of filament lamps) by using the phase-triggered power control principles described in Chapter 1, in which the triac is turned on and off once in each mains half-cycle, its M/S-ratio controlling the mean power fed to the lamp. All such circuits require the use of a simple L–C filter in the lamp feed line, to minimize RFI problems.

The three most popular ways of obtaining variable phase-delay triac triggering are to use either a diac plus *R–C* phase delay network, or to use a line-synchronized variable-delay UJT trigger, or to use a special-purpose IC as the triac trigger. *Figure 4.23* to *4.26* show practical examples of lamp dimmers using each of these three different methods.

Figure 4.23 shows a practical diac-triggered lamp dimmer, in

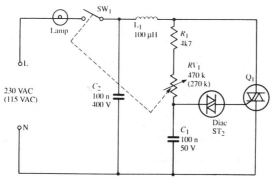

Figure 4.23 *Practical circuit of a simple lamp dimmer*

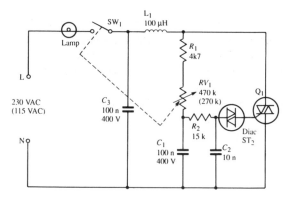

Figure 4.24   *Improved lamp dimmer with gate slaving*

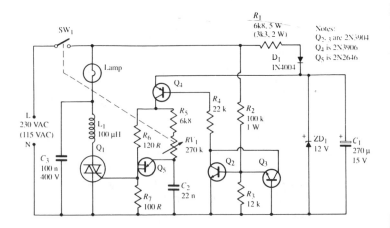

Figure 4.25   *UJT-triggered zero-backlash lamp dimmer*

which R_1–RV_1–C_1 provide the variable phase-delay. This circuit is similar to the basic lamp dimmer circuit described in Chapter 1, except for the addition of on/off switch SW_1, which is ganged to RV_1 and enables the lamp to be turned fully off.

A weakness of this simple design is that it has considerable control hysteresis or backlash, e.g., if the lamp is dimmed off by increasing the RV_1 value to 470 k, it will not go on again until RV_1 is reduced to about 400 k, and then burns at a fairly high brightness level. This backlash is caused by the diac partially discharging C_1 each time the triac fires.

The backlash effect can be reduced by wiring a 47R resistor in series with the diac, to reduce its discharge effect on C_1. An even better solution is to use the gate slaving circuit of *Figure 4.24*, in which the diac is triggered from C_2 which 'follows' the C_1 phase-delay voltage but protects C_1 from discharging when the diac fires.

If absolutely zero backlash is needed, the UJT-triggered circuit of *Figure 4.25* can be used. The UJT is powered from a 12 V DC supply derived from the AC line via R_1–D_1–ZD_1–C_1 and is line-synchronized via the Q_2–Q_3–Q_4 zero-crossing detector network, the action being such that Q_4 is turned on (applying power to the UJT) at all times other than when the AC mains voltage is close to the zero-crossover point at the end and start of each AC half-cycle. Thus, shortly after the start of each half-cycle, power is applied to the UJT circuit via Q_4, and some time later (determined by R_5–RV_1–C_2) a trigger pulse is applied to the triac gate via Q_5. The UJT resets at the end of each half-cycle, and a new sequence then begins.

A 'smart' lamp dimmer

To complete this brief look at AC lamp dimmer circuits, *Figure 4.26* shows how a dedicated IC, the Seimens S566B 'touch dimmer'

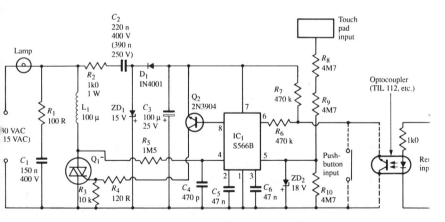

Figure 4.26 *'Smart' lamp dimmer, controlled by a dedicated IC*

chip, can be used as a 'smart' lamp dimmer that can be controlled via either touch pads, push button switches, or via an infra-red link. The action of this IC, which gives a phase-delayed trigger output to the triac, is such that it alternately ramps up (increases brilliance) or ramps down (decreases brilliance) on alternate operations of the touch or push-button inputs, but 'remembers' and holds brilliance levels when the inputs are released.

The S566B IC incorporates 'touch conditioning' circuitry, such that a very brief touch or push input causes the lamp to simply change state (from off to its previously-set on value, or vice versa), but a sustained (greater than 400 mS) input causes the IC to go into the ramping mode, in which the lamp power slowly ramps up from 3 to 97 per cent of maximum and then down to 3 per cent again, and so on until the input is released, at which point the brilliance level is latched into the dimmer's memory and used as its 'on' brilliance value.

Note that the 'touch pads' used with this circuit can be simple strips of conductive material, and that the operator is safely insulated from the AC mains voltage via high-value resistors R_8 and R_9; several touch pads or operating push buttons can be wired in parallel if desired, enabling the dimmer to be operated from several differently-placed points.

Universal-motor controllers

Domestic appliances such as electric drills and sanders, sewing machines and food mixers, etc., are almost invariably powered by series-wound 'universal' electric motors (so called because they can operate from either AC or DC supplies). When operating, these motors produce a back-e.m.f. that is proportional to the motor speed. The *effective* voltage applied to such motors is equal to the true applied voltage minus the motor's back-e.m.f. which is directly proportional to the motor speed. This fact results in a degree of self-regulation of universal motor speed, since any increase in motor loading tends to reduce the speed and back-e.m.f., thereby increasing the effective applied voltage and causing the motor speed to rise towards its original value.

Most universal motors are designed to give single-speed operation. Triac phase-controlled circuits can easily be used to provide these motors with fully-variable speed control, but give rather poor self-regulation under variable loading conditions. A suitable 'diac plus phase-delay' circuit is shown in *Figure 4.27*. This circuit is particularly useful for controlling lightly-loaded appliances such as food mixers and sewing machines, etc.

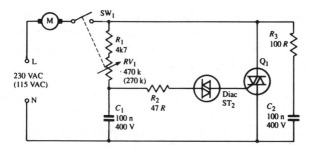

Figure 4.27 *Universal-motor speed controller, for use with lightly-loaded appliances (food mixers, sewing machines, etc.)*

Electric drills and sanders, etc., are subject to very heavy load variations, and are not suitable for control via the *Figure 4.27* circuit. Instead, the variable speed-regulator circuit of *Figure 4.28*

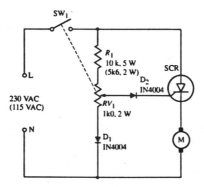

Figure 4.28 *Self-regulating universal-motor speed controller, for use with electric drills and sanders, etc.*

should be used. This circuit uses a SCR as the control element and feeds half-wave power to the motor (this results in only a 20 per cent or so reduction in maximum available speed/power), but in the off half-cycles the back-e.m.f. of the motor is sensed by the SCR and used to give automatic adjustment of the next gating pulse, to give automatic speed regulation. The $R_1 - RV_1 - D_1$ network provides only 90° of phase adjustment, so all motor pulses have minimum durations of 90° and provides very high torque. At low speeds the circuit goes into a 'skip cycling' mode, in which power pulses are provided intermittently, to suit motor loading conditions. The circuit provides particularly high torque under low-speed conditions.

5 DC power control circuits

The DC power feed to loads such as lamps, heaters, buzzers and bells, etc., can easily be controlled via unidirectional solid-state devices such as bipolar transistors, power MOSFETs, or SCRs, which can be used in either the 'static' mode to give a simple on or off switch-control action, or in the pulsed mode to give a variable power control action (as already described in Chapter 1) or a simple sound-generation action. In this chapter we show a variety of ways of using these devices to control such loads. Note that DC motor control circuits are separately described in Chapter 6.

Power-switching basics

Three basic types of solid-state device are available (in either discrete or integrated-circuit form) for use in DC power control applications, these being the bipolar transistor, the power MOSFET, and the SCR. Each of these devices offers its own particular set of advantages and disadvantages. Let us look first at the ordinary bipolar transistor.

In most DC power switching applications the bipolar transistor is wired in the common emitter mode, as shown in *Figure 5.1* and *5.2*. In the case of the npn device (*Figure 5.1*), the load is wired between Q_1 collector and the positive supply rail, and the transistor acts as a current 'sink' (current flows *into* the collector via the load). In the case of the pnp device (*Figure 5.2*), the load is wired between Q_1 collector and the negative supply rail, and the transistor acts as a current 'source' (current flows *from* the collector into the load).

The main advantage of the common emitter configuration is that

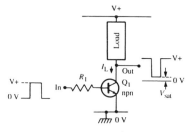

Figure 5.1 *The npn transistor acts as a load-current 'sink'*

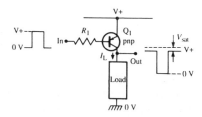

Figure 5.2 *The pnp transistor acts as a load-current 'source'*

it offers a very low saturation or 'loss' voltage (typically 200 mV to 400 mV); the main disadvantage is that it offers fairly low overall current and power gains (typically 100:1). In practice, these gains can easily be increased to 10,000:1 (without increasing the saturation voltage) by either cascading a couple of common emitter stages, as in *Figure 5.3*, or by wiring a pair of transistors in the super-alpha mode, as shown in *Figure 5.4*.

If really high current/power gains are required, they can be obtained by using a power MOSFET as the DC power switch. These devices have a near-infinite DC input impedance, and thus

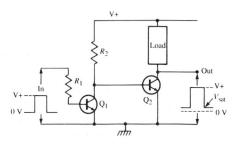

Figure 5.3 *High-gain transistor switch using cascaded npn common emitter stages*

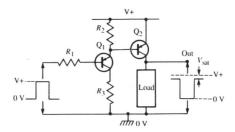

Figure 5.4   *High-gain transistor switch using a super-alpha pnp pair*

give near-infinite current and power gains. One of the most popular types of power MOSFET is the VFET family of enhancement mode devices from Siliconix, and *Figure 5.5* shows one of these devices, the VN66AF (which incorporates a built-in input-protection zener diode) used in this mode.

Note that the VN66AF can pass maximum currents of about 2 A, and has a typical saturation resistance of about 2 Ω. Thus, if the device is used with a 12 V supply, it will give a saturation voltage of about 80 mV when used with a 300 Ω load, or 1.091 V when used with a 20 Ω load, etc.

The third type of solid-state power switching device is the SCR, which is of particular value in controlling self-interrupting DC loads such as bells, buzzers or sirens. *Figure 5.6* shows the basic circuit. Usually, these loads comprise a solenoid and an activating switch wired in series to give an auto-switching action in which the solenoid first shoots forward via the close switch, and in doing so forces the switch to open, thus making the solenoid fall back and reclose the switch, thus restarting the action, and so on.

Thus, the basic *Figure 5.6* SCR circuit effectively gives a non-

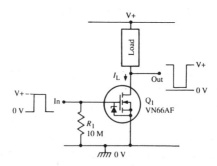

Figure 5.5   *VMOS FETs offer near-infinite current/power gain*

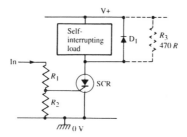

Figure 5.6   *The SCR DC switch offers high power gain*

latching load-driving action, since the SCR automatically unlatches each time the load self-interrupts. The load and SCR are thus active only while gate current is applied to the SCR. The circuit can be made fully self-latching, if desired, by shunting the load with resistor R_3, as shown, so that the SCR anode current does not fall below the SCR's minimum holding value as the load self-interrupts. Note that SCRs offer typical gate-to-anode current gains of about 5000:1, but give saturation voltage values of about 800 mV.

Load-type basics

When designing DC switching circuits, some thought must be given to the type of load to be controlled and to its possible harmful effects on the solid-state switching circuitry. The most important points to note here are as follows.

When driving filament lamps, note that these devices have a cold or 'switch-on' resistance that is typically one quarter of the normal hot or 'running' value, and thus pass switch-on currents that are four times greater than the normal running value. Thus, a solid-state switch used for controlling a 500 mA lamp must have a surge rating of at least 2 A.

When driving highly inductive loads such as relays, solenoids, bells, buzzers, speakers and electric motors, etc., note that these devices can generate very large back-e.m.f.s. at the moment of current switch-off, and solid-state power switches need protection against damage from this source. *Figure 5.7* shows how full protection can be provided by using ordinary silicon diodes to 'damp' the back-e.m.f. Here, D_1 stops the voltage from swinging more than a few hundred millivolts above the positive supply rail value, and D_2 stops it from swinging significantly below the zero

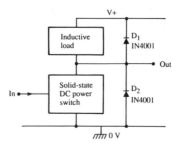

Figure 5.7 *When driving inductive loads, the solid-state switch must be protected via damping diodes*

volts rail. In practice, it is often adequate to provide only partial protection, by using only D_1, as in the case of the SCR circuit of *Figure 5.6*.

Finally, when driving loads that are electrically very noisy (such as bells, buzzers, and electric motors, etc.), note that the loads may require damping via small ceramic capacitors, to minimize RFI generation, and that the power supply to the switch-driving circuitry may need extensive ripple decoupling.

LED 'flasher' circuits

Simple 'static' DC on/off lamp or LED power switching actions are usually best carried out via conventional electromechanical switches or relays, which give minimal voltage loss but generate slight switching RFI. Repetitive DC switching, such as in lamp or LED flasher (pulser) circuits, however, is often best carried out via solid-state power switches, to minimize RFI problems, and *Figures 5.8* to *5.13* show a variety of practical circuits of this type.

Figure 5.8 shows a transistor 2-LED flasher circuit that operates

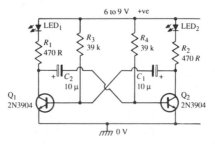

Figure 5.8 *Transistor 2-LED flasher circuit*

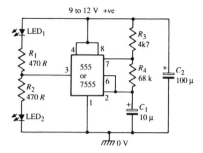

Figure 5.9 *IC 2-LED flasher circuit*

at about 1 flash-per-second and turns one LED on as the other turns off, and vice versa. The circuit is wired as an astable multivibrator, with its timing controlled via C_1–R_3 and C_2–R_4, and the LEDs and their current-limiters (R_1 and R_2) are used as the transistor collector loads. This circuit can be converted to single-LED operation by replacing the unwanted LED with a short circuit.

Figure 5.9 shows an IC version of the 2-LED flasher based on the faithful old 555 timer chip or its more modern CMOS counterpart, the 7555. The IC is wired in the astable mode, with its time constant determined by C_1 and R_4, and the action is such that IC output pin 3 alternately switches between the ground and positive supply voltage levels, alternately shorting out and disabling one or other of the two LEDs. The circuit can be converted to single-LED operation by omitting the unwanted LED and its associated current-limiting resistor.

Lamp flasher circuits

The above two circuits are suitable for driving LED loads up to only a few tens of milliamperes; if desired, greater output currents can be obtained by interposing a suitable transistor 'booster' stage between the astable output and the external load, as shown in *Figure 5.10* to *5.13*. In each of these circuits the astable is designed around one half of an inexpensive 4001B CMOS IC, and Q_2 can drive a 12 V lamp load at maximum currents up to 2 A.

In *Figure 5.10*, the astable operates at a fixed rate set by R_1 and non-polarized (NP) capacitor C_1, and turns the lamp on and off about 40 times per minute (i.e., at 1.5 seconds per flash). The astable output (from pin 4) switches alternately between 0 V and

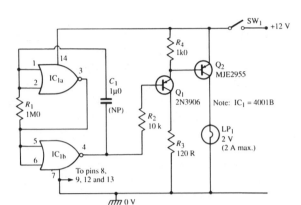

Figure 5.10 *Simple DC lamp flasher*

the full positive supply voltage; when this output is at 0 V, Q_1–Q_2 are driven on via R_2, and the lamp is illuminated, but when the output is fully positive Q_1–Q_2 are cut off, and the lamp is also off. Note that this circuit drives a lamp that has one side tied to the zero-volts rail.

The basic *Figure 5.10* circuit can be usefully modified in a number of ways, as shown in *Figure 5.11*. First, the astable frequency can be

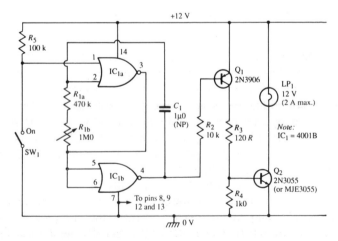

Figure 5.11 *Modified DC lamp flasher circuit*

made variable by replacing R_1 with a series-connected 470k resistor and a variable 1M0 resistor, thus making the flash rate variable between 27 and 80 times per minute. Second, the output can be used to drive lamps that have one side tied to the positive supply rail by using the alternative output stage shown in the diagram. Finally, the astable can, if desired, be turned on and off via SW_1 and the high-impedance pin-1 'gate' pin of the IC_{1a}, as shown, rather than via a switch wired in series with the positive supply rail.

The above two astable circuits generate 1:1 duty cycles or m/s-ratios, and thus turn the lamp on and off for equal periods. *Figure 5.12* shows how the basic circuit can be modified to give a

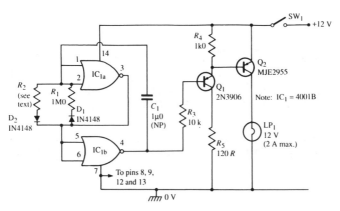

Figure 5.12   *Programmed-duty-cycle (PDC) lamp flasher circuit*

programmed duty cycle (PDC) so that, for example, the lamp gives a 0.75 seconds flash once every 8.25 seconds, thus giving a 1:10 duty cycle and giving great current economy as an emergency lamp flasher. The lamp's on time is controlled by D_1–R_1 and is fixed at about 0.75 seconds, but the off time is controlled by D_2–R_2 and can be varied over a wide range. The R_2 value can vary from a few kilohms to tens of megohms, and gives an off time of 0.75 seconds at 1M0, or 7.5 seconds at 10 MΩ, etc.

Figure 5.13 shows how the above circuit can be modified for use with the alternative type of output stage, and with direct astable on/off gating. This diagram also shows how R_2 can be replaced by a fixed and a variable resistor in series, so that the off time of the lamp is made variable. No specific values are shown for these two

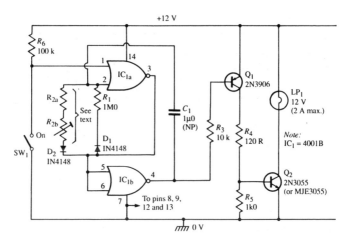

Figure 5.13 *Gated and modified PDC lamp flasher*

resistors, but note that R_{2a} determines the minimum value of the off time, and R_{2b} determines the maximum off value.

Automatic flashers

Each of the above four lamp flasher circuits are manually activated. *Figures 5.14* and *5.15* show how the basic and the PCD circuits can

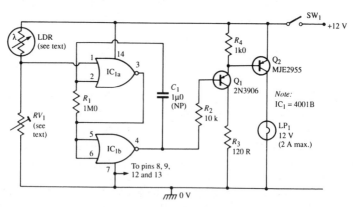

Figure 5.14 *Automatic (dark-activated) DC lamp flasher*

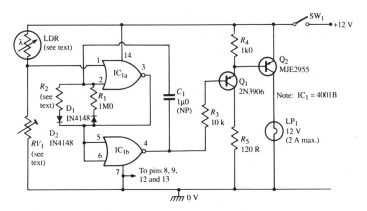

Figure 5.15 *Automatic PDC lamp flasher*

be modified so that they activate automatically when the ambient light level falls below a pre-set value.

Both of these circuits operate in the same basic way; in each case, the astable is wired in the 'gated' mode, but has its gate (pin-1) signal applied in the form of a variable voltage that is generated via light-sensitive potential divider LDR–RV_1. The action of the astable is such that it is disabled, with its pin-4 output driven high, when this gate voltage exceeds a 'threshold' value of approximately half-supply volts, but is automatically enabled when this voltage is below the threshold value.

In these circuits, the light-sensitive LDR generates a low resistance under bright conditions, and a high resistance under dark ones, and the LDR-RV_1 divider thus generates a high voltage when it is bright, and a low voltage when it is dark. In practice, RV_1 enables this voltage to be adjusted so that it precisely matches the IC's 'threshold' value at the desired light 'trip' level. Consequently, when RV_1 is suitably adjusted, these lamp flasher circuits turn on automatically when the light level falls below the desired trip level, and turn off again when it rises above that level.

The LDRs used in these circuits can be any cadmium-sulphide photocells with resistances greater than a few thousand ohms at the required turn-on light levels. RV_1 should have a maximum value roughly double that of the LDR under the turn-on condition. When building these circuits, note that the LDR faces must be shielded from the light of the flasher lamps, so that the LDRs respond to ambient light level but are unaffected by the lamp-flashing action.

Also note that master on/off switch SW_1 is wired in series with the supply line of each circuit, so that each circuit can be disabled when put away in a dark storage area.

One-shot lamp drivers

Another useful type of DC lamp-driving power control circuit is the one-shot or auto-turn-off design, which turns the lamp on as soon as a start button is pressed, but then turns it off again automatically after a preset interval variable from a few seconds to several minutes. *Figure 5.16* and *5.17* show practical circuits of this type.

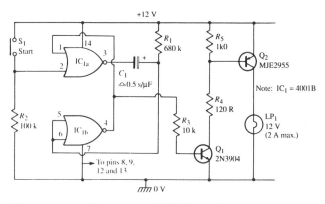

Figure 5.16 *Auto-turn-off time-controlled DC lamp driver*

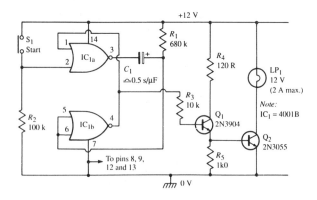

Figure 5.17 *Alternative auto-turn-off lamp driver circuit*

Both these circuits operate in the same basic way, and have two gates of a 4001B CMOS IC wired as a simple manually-triggered monostable multivibrator. Normally, the monostable output is low, so Q_1 and Q_2 are both cut off and zero power is fed to the lamp. When S_1 is briefly closed the monostable triggers and its output switches high, driving the transistors and lamp fully on. After a preset delay, the monostable output automatically switches back to the low state again, and the transistors and lamp turn off again. The circuit action is then complete.

The time delay of each circuit is determined by the C_1–R_1 values, and equals roughly 0.5 s/μF of C_1 value when R_1 equals 680k. C_1 can be an electrolytic component and can have any value up to a maximum of about 1000 μF, thus giving time delays up to several minutes; this capacitor must, however, have reasonably low leakage characteristics.

Note that these two circuits can give control of lamps that have one side taken to either the positive or the zero voltage rail, but that the transistor output stages differ from those described earlier.

DC lamp-dimmer circuits

It was explained in Chapter 1 that a load's DC power levels can be varied by using either 'variable-voltage' or 'switched-mode' control techniques, and either of these techniques can in fact be used to control the brightness of a DC filament lamp. *Figure 5.18* shows the practical circuit of a simple variable-voltage type of DC lamp dimmer.

In *Figure 5.18*, RV_1 acts as a variable potential divider which applies an input voltage to the base of emitter follower Q_1, which buffers (power boosts) this voltage and applies it to the lamp. RV_1 thus enables the lamp voltage (and thus its brilliance) to be fully

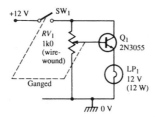

Figure 5.18 *'Variable-voltage' DC lamp brightness control circuit*

varied from zero to maximum. Disadvantages of this simple circuit are that it is very inefficient, since all unwanted power is 'lost' across Q_1, and Q_1 needs a large power rating and must be capable of handling the cold current of the lamp.

Figure 5.19 illustrates the basic principles of switched-mode variable power control. Here, an electronic switch (SW_1) is wired in series with the lamp and can be opened and closed via a pulse-generator waveform. When this pulse is high, SW_1 is closed and power is fed to the lamp; when the pulse is low SW_1 is open, and power is not fed to the lamp.

The important thing to note about the *Figure 5.19* pulse generator is that it generates a waveform with a fixed frame width but with a variable M/S-(on/off) ratio or duty cycle, thereby enabling the *mean* lamp voltage to be varied. Typically, the M/S-ratio is fully variable from 1:20 to 20:1, enabling the mean lamp voltage to be varied from 5 to 95 per cent of the DC supply-voltage value.

Because of the inherently long thermal time constant of a filament lamp, its brilliance responds relatively slowly to rapid changes in input power. Consequently, if the frame width of the *Figure 5.19* waveform generator is less than roughly 20 mS (i.e., the repetition frequency is greater than 50 Hz), the lamp will show no sign of flicker, and its brilliance can be varied by altering the M/S-ratio.

Thus, if the M/S-ratio of the *Figure 5.19* circuit is set at 20:1, the mean lamp voltage is 11.4 V and the consequently hot lamp consumes 10.83 W. Alternatively, with the M/S-ratio set at 1:20, the mean lamp voltage is only 600 mV, so the lamp is virtually cold and consumes a mere 120 mW. The lamp power consumption can thus be varied over a 90:1 range via the M/S-ratio control. Note, however, that this wide range of control is obtained with virtually zero power loss within the system, since power is actually controlled by SW_1, which is always either fully on or fully off. The

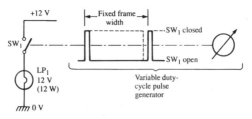

Figure 5.19   *Basic 'switched-mode' DC lamp brightness control circuit*

switched-mode control system is thus highly efficient (typical efficiency is about 95 per cent).

Figure 5.20 shows a practical switched-mode DC lamp dimmer circuit. Here, IC_{1a} and IC_{1b} are wired as an astable multivibrator that operates at a fixed frequency of about 100 Hz and has its output fed to the lamp via Q_1 and Q_2, but has the mark part of its waveform controlled via C_1–D_1–R_1 and the right-hand part of RV_1, and the *space* part controlled via C_1–D_2–R_2 and the left-hand part of RV_1, thus enabling the M/S-ratio to be varied from 1:20 to 20:1 via RV_1, thus enabling the mean lamp power to be varied over a 90:1 range. Note that on/off switch SW_1 is ganged to RV_1, enabling the circuit to be turned fully off by turning the RV_1 brilliance control fully anticlockwise, and that R_6–C_2 protect IC_1 against damage from supply-line transients, thus enabling the circuit to be powered via 12 V motor vehicle supplies, etc.

The *Figure 5.20* circuit can be used to control the brilliance of any low power (up to 24 W) 12 V filament lamp that has one side wired to the zero volt rail of the power supply. *Figure 5.21* shows how the design can be modified (by using an alternative output stage) for use with lamps that have one side wired to the positive rail of the power supply.

Finally, *Figure 5.22* shows how the basic switched-mode circuit can be used to efficiently control the brilliance of one (or more) LEDs, at maximum currents up to about 20 mA. In this case the two spare gates of IC_1 are wired as simple parallel-connected inverters and used to provide a medium-current buffered drive to

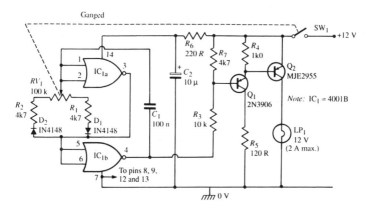

Figure 5.20   *Switched-mode DC lamp dimmer*

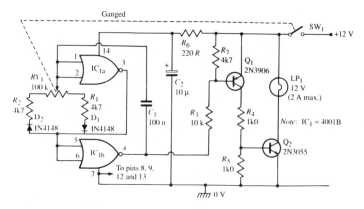

Figure 5.21　*Alternative switched-mode DC lamp dimmer*

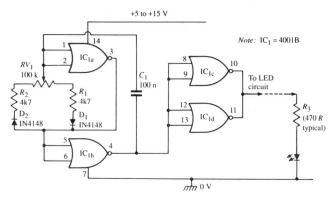

Figure 5.22　*Switched-mode LED brightness-control circuit*

the LED via current-limiter R_3. This circuit can be powered from any DC supply in the range 5 to 15 V.

Bell/buzzer alarm circuits

One of the easiest ways of activating a self-interrupting audible alarm device such as a bell or buzzer is via a SCR, and *Figures 5.23* to *5.31* show a selection of circuits of this type. All of these are designed around the inexpensive and readily available type C106 SCR, which can handle mean load currents up to 2.5 A, needs a

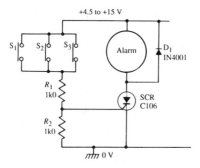

Figure 5.23 *Multi-input non-latching alarm circuit*

gate current of less than 200 μA, and has a 'minimum holding current' value of less than 3 mA. Note in all cases that the circuit's supply voltage should be about 1.5 V greater than the nominal operating voltage of the alarm device used, to compensate for voltage losses across the SCR, and that diode D_1 is used to damp the alarm's back-e.m.f.s.

Figure 5.23 shows the circuit of a simple non-latching multi-input alarm, in which the alarm activates when any of the S_1 to S_3 push-button input switches is closed, but stops operating as soon as the switch is released.

Figure 5.24 shows how the above circuit can be converted into a self-latching multi-input 'panic' alarm (which can be activated by owners if they feel immediately threatened while at home) by simply wiring R_3 plus normally-closed *reset* switch S_4 in parallel with the alarm device, so that the SCR's main-terminal current

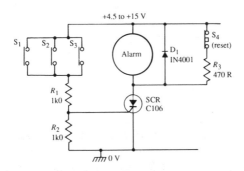

Figure 5.24 *Multi-input self-latching 'panic' alarm*

does not fall below the 3 mA 'minimum holding' value when the alarm self-interrupts. Once this circuit has latched, it can be unlatched again (reset) by briefly opening S_4.

Note that both of the above circuits pass typical standby currents of only 0.1 μA when the alarm is in the off condition, and can thus be powered from battery supplies, and that the S_1 to S_3 switches pass activating currents of only a few milliamperes, and can thus safely be connected to the alarm circuit via considerable lengths of cable.

Figure 5.25 shows how the self-latching circuit can be converted into a simple burglar alarm system, complete with the 'panic' facility. Here, the alarm can be activated by briefly opening any of the series-connected normally-closed S_1 to S_3 'burglar alarm' switches (which can take the form of microswitches that are activated by the action of opening doors or windows, etc.), or by briefly closing any of the parallel-connected normally-open 'panic' switches. Note that this circuit passes a typical standby current of 0.5 mA (via R_1) when powered from a 6 V supply, and that C_1 acts as a noise-suppressing capacitor that ensures that the alarm will only operate if the S_1 to S_3 switches are held open for more than a millisecond or so, thus enhancing the circuit's reliability.

The standby current of the burglar alarm circuit can be reduced to a mere 1.4 μA (at 6 V) by modifying it as shown in *Figure 5.26*, where Q_1 and Q_2 are connected in the Darlington mode and wired as a common-emitter amplifier, which inverts and boosts the R_1-derived 'burglar' signal and then passes it on to the gate of the SCR.

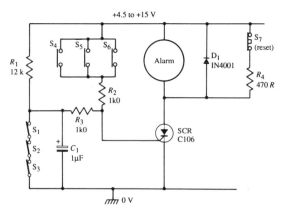

Figure 5.25   *Simple burglar alarm system, with 'panic' facility*

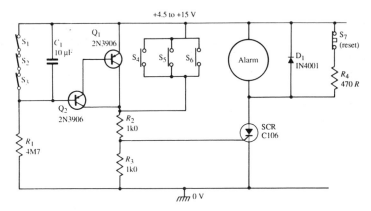

Figure 5.26 *Improved burglar alarm circuit*

Water, light and heat alarms

The basic SCR-driven alarm circuit can be used to indicate the presence of excess water, light, or temperature levels by driving the SCR gate via suitable 'detection' circuitry. *Figures 5.27* to *5.31* show alarm circuits of this type.

The *Figure 5.27* 'water-activated' alarm uses Q_1 to effectively increase the SCR's gate sensitivity, and activates when a resistance of less than about 220 kΩ appears across the two metal probes. Its operation as a water-activated alarm relies on the fact that the impurities in normal water (and many other liquids and vapours) make it act as a conductive medium with a moderately low electrical resistance, which thus causes the alarm to activate when water comes into contact with both probes simultaneously. C_1 is

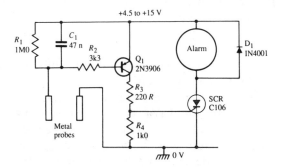

Figure 5.27 *Water-activated alarm*

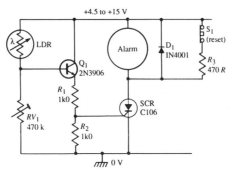

Figure 5.28 *Light-activated alarm*

used to suppress unwanted AC signal pick-up, and R_2 limits Q_1's base current to a safe value. By suitably adjusting the placing of the two metal probes, this circuit can be used to sound an alarm when water rises above a preset level in a bath, tank, or cistern, etc.

The operation of the *Figure 5.28* 'light-activated' alarm is very simple. The LDR and RV_1 are wired as a light-sensitive voltage-generating potential divider that has its output buffered via Q_1 and fed to the SCR gate via R_1. This output is low under dark conditions (when the LDR resistance is high), but goes high under bright conditions (when the LDR resistance is low) and thus drives the SCR and alarm on. The precise light-triggering point of the circuit can be pre-set via RV_1, and almost any small cadmium sulphide photocell can be used in the LDR position. This circuit can be used to sound an alarm when light enters a normally-dark area such as a drawer or wall safe, etc.

Temperature-activated alarms can be used to indicate either fire or overheat conditions, or frost or underheat conditions. *Figures 5.29* to *5.31* show three such circuits; in each of these designs TH_1 can be any NTC thermistor that presents a resistance in the range 1k0 to 20k at the required trigger temperature, and preset pot RV_1 should have a maximum resistance value roughly double that of TH_1 under this trigger condition.

The *Figure 5.29* over-temperature alarm operates as follows. R_1–R_2 and TH_1–RV_1 are wired as a Wheatstone bridge in which R_1–R_2 generates a fixed half-supply 'reference' voltage and TH_1–RV_1 generates a temperature-sensitive 'variable' voltage, and Q_1 is used as a bridge balance detector and SCR gate driver. RV_1 is adjusted so that the bridge is normally balanced, with the 'reference' and

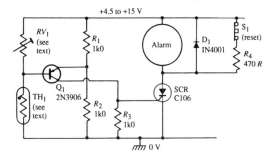

Figure 5.29 *Simple overtemperature alarm*

'variable' voltages at equal values, at a temperature just below the required trigger value, and under this condition Q_1 base and emitter are at equal voltages and Q_1 and the SCR are thus cut off. When the TH_1 temperature is below this 'balance' value the TH_1–RV_1 voltage is above the 'reference' value, so Q_1 is reverse biased and the SCR remains off, but when the TH_1 temperature is significantly above the 'balance' value the TH_1–RV_1 voltage is below the 'reference' value, so Q_1 is forward biased and drives the SCR on, thus sounding the alarm. The precise trigger point of the circuit can be preset via RV_1.

The action of the above circuit can be reversed, so that the alarm turns on when the temperature falls below a preset level, by simply transposing the TH_1 and RV_1 positions, as shown in the frost or under-temperature alarm circuit of *Figure 5.30*.

The above two circuits perform very well, but their precise trigger points are subject to slight variation with changes in Q_1 temperature, due to the temperature dependence of the Q_1 base-emitter junction characteristics. These circuits are thus not suitable

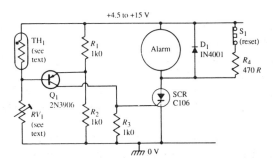

Figure 5.30 *Simple frost or undertemperature alarm*

for use in precision applications, unless Q_1 and TH_1 operate at equal temperatures. This snag can be overcome by using a two-transistor differential detector in place of Q_1, as shown in *Figure 5.31*.

The *Figure 5.31* circuit is wired as a precision over-temperature alarm. It can be made to function as a precision under-temperature alarm by simply transposing the RV_1 and TH_1 positions. Note that the circuit is shown without a latching resistor, since sensitive circuits of this type are usually required to sound the alarm only so long as the TH_1 temperature is beyond the preset limit.

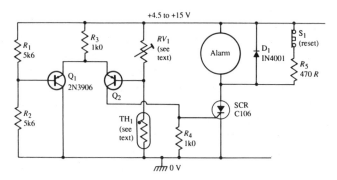

Figure 5.31   *Precision overtemperature alarm*

Piezoelectric alarms

Piezoelectric transducers are widely used as sound generators in gadgets such as toys, clocks, watches, calculators, and electronic games, etc., and in a variety of applications where space and operating efficiency are at a premium. They consist of a thin slice of electroconstrictive (piezo) ceramic material plus two electrical contacts, and act as super-efficient electric-to-acoustic power converters when operated in the 1 kHz to 5 kHz frequency range. They give typical power conversion efficiencies of 50 per cent, compared to about 0.5 per cent for conventional loudspeakers, and thus act as excellent 'sound makers'.

Piezoelectric transducers are available from several manufacturers. The Toko PB2720 is fairly typical of the type. It houses the actual transducer in a small 'easy-to-use' plastic-moulded housing, and its two input terminals appear to the outside world as a simple capacitor with a static value of about 20 nF and a DC resistance of near-infinity. The most effective and cheapest way to drive the device is to feed it with square waves, but in this case the driver must be able to source and sink currents with equal ease and must have a current-limited (short-circuit proof) output. CMOS drivers fit this bill perfectly.

Figure 5.32 and *5.33* show two inexpensive ways of driving the PB2720 (or any similar device) from a 4011B CMOS astable oscillator. Each of these circuits generates a 2 kHz monotone signal when in the on mode, is gated on by a high (logic–1) input, and can use any DC supply in the range 3 to 15 V.

In the *Figure 5.32* design, IC_{1a}–IC_{1b} are wired as a 2 kHz astable that can be gated on electronically or via push-button switch S_1, and IC_{1c} is used as an inverting buffer/amplifier that gives single-ended drive to the PB2720. The signal reaching the PB2720 is thus a square wave with a peak-to-peak amplitude roughly equal to the supply voltage, and the r.m.s. signal voltage across the load equals roughly 50 per cent of the supply line value.

The *Figure 5.33* design is similar to the above, except that inverting amplifiers IC_{1c} and IC_{1d} are series-connected and used to give a 'bridge' drive to the transducer, with antiphase signals being fed to the two sides of the PB2720. The consequence of this drive technique is that the load (the PB2720) actually sees a square wave drive voltage with a peak-to-peak value equal to twice the supply voltage value, and an r.m.s. voltage equal to the supply value, and thus gives four times more acoustic output power than the *Figure*

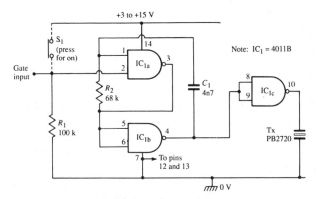

Figure 5.32 *Gated piezoelectric monotone alarm with single-ended output*

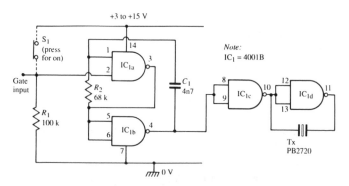

Figure 5.33   *Gated piezoelectric monotone alarm with bridge-drive output*

5.32 design. This action can be understood with the aid of *Figure 5.34*, which shows the waveforms applied to the load from the 'bridge' circuit when it is fed with a 10 V peak-to-peak square wave input signal.

Note in *Figure 5.34* that although waveforms A and B each have peak values of 10 V relative to ground, the two signals are in antiphase (shifted by 180°). Thus, during period 1 of the drive signal, point B is 10 V positive to A and is thus seen as being at +10 V. In period 2, however, point B is 10 V negative to point A, and is

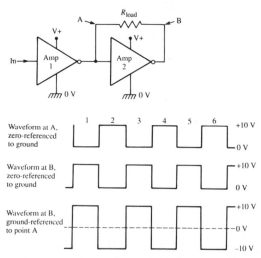

Figure 5.34   *A pair of amplifiers connected in the 'bridge-driving' mode give a power output of $2V^2/R$ W, i.e., four times the power of a single-ended output*

seen as being at −10 V. Consequently, if point A is regarded as a zero voltage reference point, it can be seen that the point B voltage varies from +10 V to −10 V between periods 1 and 2, giving a total voltage change of 20 V across the load. Similar changes occur in all subsequent waveform periods.

Thus, the load in a 10 V bridge-driven circuit sees a total voltage of 20 V peak-to-peak, or twice the single-ended input voltage value, as indicated in the diagram. Since doubling the drive voltage results in a doubling of drive current, and power equals the *V–I* product, the bridge-driven circuit thus produces four times more power output than a single-ended circuit.

Alarm circuit variations

Gated CMOS oscillator/driver circuits can be used in a variety of ways to produce useful sounds from the PB2720. A few variations are shown in *Figure 5.35* to *5.37*. *Figure 5.35*, for example, shows how the basic bridge-driving circuit can be modified so that it can be gated on by a low (logic 0) input (rather than a high one) by simply substituting a 4001B CMOS IC for the 4011B type.

Figure 5.36 shows how to use a single 4011B to make a pulsed-tone (bleep-bleep) alarm circuit with direct drive to the PB2720. Here, IC_{1a}–IC_{1b} are wired as a gated 6 Hz astable and is used to gate the IC_{1c}–IC_{1d} 2 kHz astable on and off. This circuit is gated on by a high input. If low-input gating is wanted, simply swap the 4011B for a 4001B and transpose the positions of S_1 and R_1.

Finally, *Figure 5.37* shows a warble-tone (dee-dah-dee-dah)

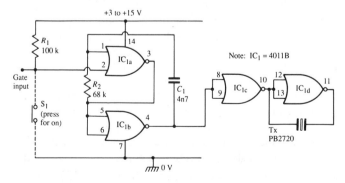

Figure 5.35 *Alternative version of the gated bridge-driving circuit*

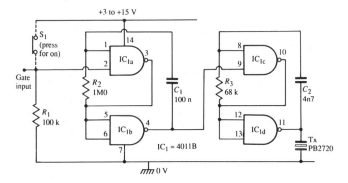

Figure 5.36   *Gated pulsed-tone (6 Hz and 2 kHz) alarm with piezo output*

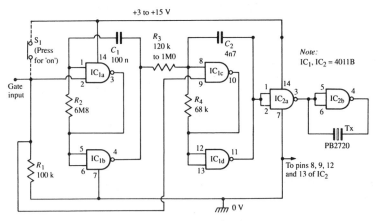

Figure 5.37   *Gated warble-tone alarm with bridge-driven piezo output*

version of the gated alarm which generates a sound similar to a British police car siren and has a bridge-driven output. Here, 1 Hz astable IC_{1a}–IC_{1b} is used to modulate the frequency of the IC_{1c}–IC_{1d} astable; the depth of frequency modulation depends on the value of R_3 which can have any value from 120 kΩ to 1 MΩ.

Loud speaker alarms

The basic CMOS alarm-sound generator astable circuits of *Figures 5.32* to *5.37* can easily be modified to generate acoustic outputs via loud speakers, thus making greater sound levels available. *Figure 5.38*, for example, shows how the *Figure 5.35* 4001B gated astable

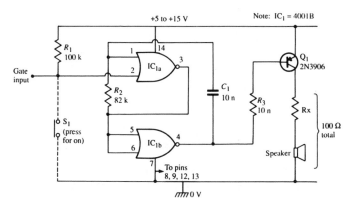

Figure 5.38 *Low-power 800 Hz monotone alarm with speaker output*

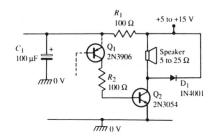

Figure 5.39 *Medium-power (0.25 W to 11.25 W) booster stage*

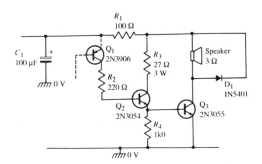

Figure 5.40 *High-power (18 W) booster stage*

can be operated at 800 Hz and used to generate such an output. Here, the astable's pin-4 output is fed to the base of pnp common emitter 'driver' amplifier Q_1 via R_3, and Q_1 uses the speaker and current-limiting resistor Rx as its collector load. The output of the 4001B astable goes high when gated off, and under this condition the pnp transistor is cut off and consumes zero current; when the astable is active, Q_1's output switches on and off at an 800 Hz rate and generates an acoustic output via the speaker.

Note that the 4011B version of the gated CMOS astable (see *Figures 5.32* and *5.33*, etc.) gives a pin-4 output that is grounded when gated off, and this output must thus be fed to the input of an npn driver stage (with a grounded emitter) if a speaker output equal to the above is required.

The basic *Figure 5.38* circuit is intended for low-power applications, and can be used with any speaker in the range 3 Ω to 100 Ω and with any supply in the range 5 to 15 V. Note that resistor Rx is wired in series with the speaker and must have its value chosen so that the total resistance is roughly 100 Ω, to keep the dissipation of Q_1 within acceptable limits. The mean output power level of the circuit depends on the individual values of speaker impedance and supply voltage used, but is usually of the order of only a few tens of milliwatts. Using a 9 V supply, for example, the output power to a 15 Ω speaker is about 25 mW, and to a 100 Ω speaker is about 160 mW.

If desired, the output power of the above circuit can be greatly increased by modifying its output to accept the power booster circuits of *Figure 5.39* or *5.40*. In these circuits, R_2 is wired in series with the collector of the existing Q_1 alarm output transistor and provides base drive to a one- or two-transistor booster stage, and the alarm's power supply is decoupled from that of the booster via R_1–C_1. Note that protection diodes are wired across the speakers of these circuits, to prevent the speaker back e.m.f.s. from exceeding the supply rail voltage.

The *Figure 5.39* booster circuit can be used with any speaker in the range 5 Ω to 25 Ω and with any supply from 5 V to 15 V. The available output power varies from 250 mW when a 25 Ω speaker is used with a 5 V supply, to 11.25 W when a 5 Ω speaker is used with a 15 V supply. The *Figure 5.40* circuit is designed to operate from a fixed 15 V supply and uses a 3 Ω speaker, and gives a mean output power of about 18 W. Note that, because of transistor leakage currents, these circuits pass typical quiescent current of about 20 μA when in the standby mode.

6 DC motor control circuits

One of the most interesting applications of electronic power control techniques is in the control of DC electric motors. These techniques can be used to give precision step rotation or speed/direction control of multiphase stepper motors, or precision speed regulation or wide-range speed control of permanent magnet DC motors, or precision control of the speed or angular movement of various types of DC servomotor, etc. Practical application circuits of all these types are described in this chapter.

Motor types

There are four major types of motor that are relevant to this chapter. The first of these is a fairly modern development, the so-called 'stepper' motor. These motors are provided with a number of phase windings, and each time these are electrically pulsed in the appropriate sequence the output spindle rotates by a precise step angle (typically between 1.8 and 7.5°). Thus, by applying a suitable sequence of pulses, the spindle can be turned a precise number of steps backwards or forwards, or can be made to rotate continuously at any desired speed in either direction.

Stepper motors can easily be controlled via a microprocessor or via dedicated stepper motor driver ICs such as the SAA1027 or SAA1024, etc., and are widely used in all applications where precise amounts of angular movement are required, such as in the movement of robot arms, in daisywheel character selection, or the movement control of the printhead and paper feed in an electronic typewriter, etc.

The most widely used type of DC motor is the permanent magnet commutator type, which is simply designed to rotate at some approximate speed when powered via a particular DC voltage. Motors of this type are often used as fixed-speed drivers in tape-cassette recorders and record/disc players, and as wide-range variable-speed drivers in miniature electric drills and model loco-motives, etc. In all of these applications, the motor performance can be greatly enhanced with the aid of electronic control circuitry.

The third type of motor is the so-called 'servomotor', which is an electric motor that is mechanically coupled to a movement-to-data translator such as a shaft-mounted tachogenerator, which gives an output directly proportional to the motor speed, or a potentiometer that is mounted on a gearbox-driven output shaft, thus giving an output that is directly proportional to the shaft position. When one of these motors is coupled into a suitable power control feedback loop, its speed can be precisely locked to that of an external frequency generator, or its shaft movements can be locked to that of an external shaft or control knob.

Servomotors of the tachogenerator type are often used to give precision speed control of record/disc turntables, etc., and servo-motor of the 'pot output' type are widely used to give remote-controlled antenna rotation or remote activation of model aircraft control surfaces and engine speed, etc.

The final type of motor is the 2-phase low-voltage AC motor, which is usually driven via a DC-powered low frequency oscillator. Motors of this type are occasionally used in turntable driving applications.

Stepper motor basics

Stepper motors come in two basic forms, being either variable-reluctance types, or hybrid types. The basic operating principle of the stepper motor can best be understood by looking at *Figure 6.1*, which shows four sequential stepping operations of a 4-phase variable-reluctance motor.

The stator (body) of this motor has eight inward-projecting teeth, each with a winding or coil that is connected in an opposing sense to the coil on the opposite tooth, to form one phase; there are a total of four phases. When a phase is energized, it generates a magnetic flux that flows from the positive phase tooth to the

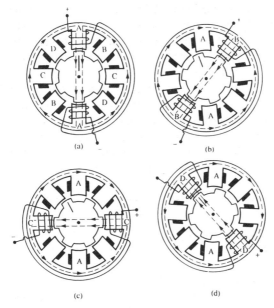

Figure 6.1 *Motor actions in four sequential stepping actions of a 4-phase variable-reluctance stepper motor*

negative one via the shortest possible magnetic path through the soft-iron rotor. The rotor has six (in this case) projecting teeth, and to minimize this magnetic path the rotor is forced to move so that the nearest pair of its teeth align with those of the energized phase.

Thus, in *Figure 6.1(a)* phase A is energized and the rotor's reference tooth (indicated by a short line) aligns with positive phase tooth A. In *Figure 6.1(b)* phase B is energized, and the rotor is forced to step 15° anticlockwise as the B-phase teeth align with the nearest set of rotor teeth. This process repeats in *Figures 6.1(c)* and *(d)* as phases C and D are sequentially energized, forcing the rotor to step 15° anticlockwise in each case, to give a total of 45° of movement after three phase steps, at which point the rotor's reference tooth has aligned with the phase-D teeth, and the whole A–B–C–D phase stepping sequence can be repeated.

Note from the above that the rotor can be moved in an anticlockwise direction by switching the phases in an A–B–C–D sequence, or clockwise by switching them in a D–C–B–A sequence, and that the motor can be spun in either direction by continuously repeating the appropriate set of sequences.

The step length of this type of motor equals $360/p.n$ degrees, where p is the number of phases, and n is the number of rotor teeth. In the case of *Figure 6.1*, this gives a step length of 15°, indicating that a 24-step sequence is needed to complete one motor revolution.

Hybrid stepper motors

The most popular variety of stepper motor is the so-called hybrid type, which gives the same type of stepping action as that of *Figure 6.1*, but differs in its form of construction and operation, in that its rotor incorporates a permanent magnet, and its energizing flux flows parallel to the axis of the shaft, etc. Usually, these motors have four phases or coil windings, which may be available via eight independent terminals, as shown in *Figure 6.2(a)*, or via two sets of triple terminals, as shown in *Figure 6.2(b)*. Note that the phases are usually designed for unipolar drive, and must be connected in the correct polarity.

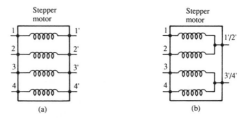

Figure 6.2   *4-phase stepper motors usually have either (a) eight or (b) six external connection terminals*

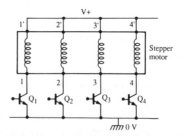

Figure 6.3   *Basic transistor-driven stepper motor circuit*

Step No.	Q_1	Q_2	Q_3	Q_4
0	On	Off	On	Off
1	Off	On	On	Off
2	Off	On	Off	On
3	On	Off	Off	On
4	On	Off	On	Off
5	Off	On	On	Off

Above sequence repeating · Clockwise / Anticlockwise

Figure 6.4 *Full-step mode sequencing of the Figure 6.3 circuit*

Step No.	Q_1	Q_2	Q_3	Q_4
0	On	Off	On	Off
1	On	Off	Off	Off
2	On	Off	Off	On
3	Off	Off	Off	On
4	Off	On	Off	On
5	Off	On	Off	Off
6	Off	On	On	Off
7	Off	Off	On	Off
8	On	Off	On	Off
9	On	Off	Off	Off

Above sequence repeating · Clockwise / Anticlockwise

Figure 6.5 *Half-step mode sequencing of the Figure 6.3 circuit*

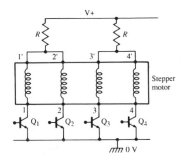

Figure 6.6 *Method of driving a stepper motor from a DC supply greater than its rated value*

Figure 6.3 shows the basic way of transistor driving a normal 4-phase hybrid stepper motor at its designated voltage rating, and *Figure 6.4* shows the normal full-step switching sequence. Note that the motor can be repeatedly stepped or rotated clockwise by repeating the 1–2–3–4 sequence or anticlockwise by repeating the 4–3–2–1 sequence. Also note that, in each step, two phases are energized at the same time, but that phases 1 and 2 or 3 and 4 are never both on at the same time.

A useful feature of the 4-phase hybrid motor is that it can also be driven in the 'half step' mode, in which the rotor advances only a half step angle at a time, by using a mixture of single and dual phase switching, as shown in *Figure 6.5*.

A 4-phase hybrid motor can be operated via a DC supply greater than its designated voltage rating by wiring suitable dropper resistors in series with its phases. Since phases 1 and 2 or 3 and 4 are never both on at the same time, however, each of these pairs of phases can share a single dropper resistor, as shown in *Figure 6.6*. Thus, a 6 V, 6 Ω motor (1 A per phase) can be operated via a 12 V supply by giving each resistor a 6 Ω, 6 W rating.

The SAA1027 driver IC

A number of dedicated 4-phase stepper motor driver ICs are available, and the best known of these is the SAA1027, which is designed to operate from supplies in the 9.5 to 18 V range and to give full-stepping 4-phase motor operation at total output drive currents up to about 500 mA.

Figure 6.7 shows the outline and pin notations of the SAA1027 IC, *Figure 6.8* shows its internal block diagram, and *Figure 6.9* shows its basic application circuit. Internally, the IC has three buffered inputs which are used to control a synchronous 2-bit (four-state) up/down counter, which has its output fed to a code converter which then uses its four outputs to control (via suitable driver circuitry) four transistor output stages which each operate in

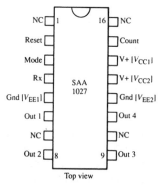

Figure 6.7   *Outline and pin designations of the SAA1027 stepper motor driver IC*

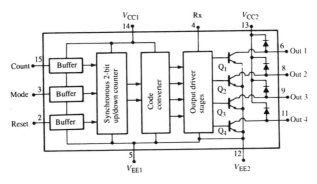

Figure 6.8 *Internal block diagram of the SAA1027*

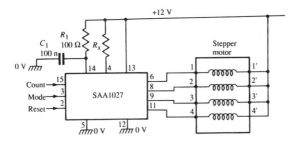

Figure 6.9 *Basic SAA1027 application circuit*

the open-collector mode and is protected against damage from motor back-e.m.f.s via an internal collector-to-pin-13 diode.

Note that the IC has two sets of supply rail pins, with one set (pins 13 and 12) feeding the high-current circuitry, and the other (pins 14 and 5) feeding the low current sections. In use, pins 5 and 12 are grounded, and the positive (usually 12 V) rail is fed directly to pin 13 and via decoupling component R_1–C_1 to pin 14. The positive rail must also be fed to pin 4 via Rx, which determines the maximum drive current capacity of the four output transistors; the appropriate Rx value is given by:

$$Rx = (4E/I) - 60 \ \Omega$$

Where E is the supply voltage and I is the desired maximum motor phase current. Thus, when using a 12 V supply, Rx needs values of 420 Ω, 180 Ω, or 78 Ω for maximum output currents of 100 mA, 200 mA, or 350 mA respectively.

The SAA1027 IC has three input control terminals, notated

Counting sequence	Mode = low				Mode = High			
	Q_1	Q_2	Q_3	Q_4	Q_1	Q_2	Q_3	Q_4
0	On	Off	On	Off	On	Off	On	Off
1	Off	On	On	Off	On	Off	Off	On
2	Off	On	Off	On	Off	On	Off	On
3	On	Off	Off	On	Off	On	On	Off
0	On	Off	On	Off	On	Off	On	Off
Reset low	On	Off	On	Off	On	Off	On	Off

Above sequence → repeating

Figure 6.10   *SAA1027 output sequencing table*

count, mode, and reset. The *reset* terminal is normally biased high, and under this condition the IC's outputs change state each time the *count* terminal transitions from the low to the high state, as shown in the output sequencing table of *Figure 6.10*. The sequence repeats at 4-step intervals, but can be reset to the zero state at any time by pulling the reset pin low. The sequence repeats in one direction (normally giving clockwise motor rotation) when the *mode* input pin is tied low, and in the other (normally giving anticlockwise motor rotation) when the mode input pin is tied high.

Figure 6.11 shows a practical drive/test circuit that can be used to activate hybrid 4-phase stepper motors with current rating up to about 300 mA. The motor can be manually sequenced one step at a time via SW_3 (which is effectively 'debounced' via R_4–C_5), or automatically via the 555/7555 astable oscillator, by moving SW_2 to either the step or free-run position; the motor direction is controlled via SW_4, and the stepping sequence can be reset via SW_5.

The operating speed of the free-running astable circuit is fully variable via RV_1, and is variable in three switch-selected decade ranges via SW_1. In the slow (1) range, the astable frequency is

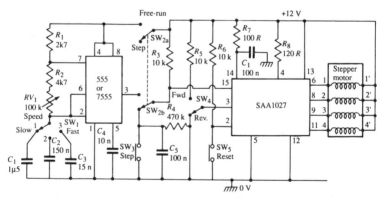

Figure 6.11   *Complete stepper motor drive/test circuit*

variable from below 5 Hz to about 68 Hz via RV_1; on a 48-step (7.5°
step angle) motor this corresponds to a speed range of 6 to 85 r.p.m.
SW_1 ranges 2 and 3 give frequency ranges that are ten and 100 times
greater than this respectively, and the circuit thus gives a total
speed control range of 6 to 8500 r.p.m. on a 48-step motor.

Circuit variations

The basic *Figure 6.11* circuit can be varied in a variety of ways.
Figure 6.12, for example, shows how it can be driven via a

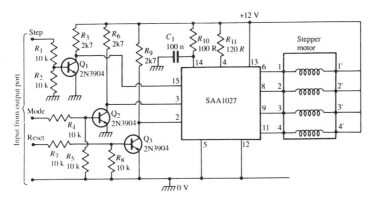

Figure 6.12 *Stepper-motor-to microprocessor interface*

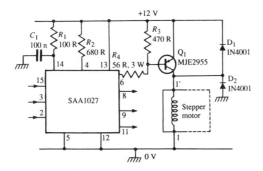

Figure 6.13 *Method of boosting the drive current to stepper motors with independent phase windings*

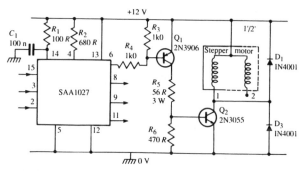

Figure 6.14 *Method of boosting the drive current to stepper motors with coupled phase windings*

microprocessor or computer output port with terminal voltages that are below 1 V in the logic-0 state and above 3.5 V in the logic-1 state. Note that this circuit reverses the normal polarity of the input control signals; thus, the step input is pulsed by a high-to-low transition, the stepping sequence is reset by a high input, and a low mode input gives forward motor rotation and a high input gives reverse rotation.

The *Figure 6.11* and *6.12* circuits are designed to give maximum output drive currents up to about 300 mA. If desired, these outputs can be boosted to a maximum of about 5 A by using the circuits of *Figure 6.13* or *6.14*, which each show the additional circuitry needed to drive one of the four output phases of the stepper motor; four such driver stages are needed per motor. The *Figure 6.13* circuit can be used to drive motors with fully independent phase windings, and the *Figure 6.14* design can be used in cases where two windings share a common supply terminal. Note in both cases that D_1 and D_2 are used to damp the motor back-e.m.f.s.

Magnet/commutator motor basics

The most widely used type of DC motor is the permanent magnet commutator type, which has a commutator that is designed to rotate when the motor is powered from an appropriate DC voltage. *Figure 6.15(a)* and *(b)* shows the circuit symbol and the simplified equivalent circuit of this type of motor.

The basic action of this motor is such that an applied DC voltage

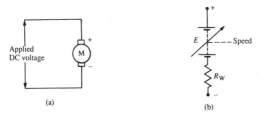

Figure 6.15 *(a) Symbol and equivalent circuit (b) of a permanent-magnet type of DC motor*

causes a current to flow through sets of armature windings (via commutator segments and pick-up brushes) and generate electromagnetic fields that react with the fields of fixed stator magnets in such a way that the armature is forced to rotate; as it rotates, its interacting fields force it to generate a back-e.m.f. that opposes the applied DC voltage and is directly proportional to the armature speed, thus giving the equivalent circuit of *Figure 6.15(b)*, in which R_w represents the total resistance of the armature windings, etc., and E represents the speed-dependent back-e.m.f. Important points to note about this kind of motor are as follows:

1　When the motor is loaded by a fixed amount, its speed is directly proportional to supply voltage.
2　When the motor is powered from a fixed DC supply, its running current is directly proportional to the amount of armature loading.
3　The motor's *effective* applied voltage equals the applied DC voltage minus the speed-dependent back-e.m.f. Consequently, when it is powered from a fixed voltage, motor speed tends to self-regulate, since any increase in loading tends to slow the armature, thus reducing the back-e.m.f. and increasing the *effective* applied voltage, and so on.
4　The motor current is greatest when the armature is stalled and the back-e.m.f. is zero, and then equals V/R_w (where V is the supply voltage); this state naturally occurs under 'start' conditions.
5　The direction of armature rotation can be reversed by reversing the motor's supply connections.

The main applications of electronic power control to DC motors of this type are in on/off switching control, in direction control, in improved speed self-regulation, and in variable speed control; all of these subjects are dealt with in the next few pages.

On/off switching

A DC motor can be turned on and off by wiring a control switch between the motor and its power supply. This switch can be an ordinary electromechanical type (or a pair of relay contacts), as in *Figure 6.16(a)*, or it can be a switching transistor, as in the *Figure 6.16(b)* circuit, in which the motor is off when the input is low and is on when the input is high. Note here that diodes D_1 and D_2 are used to damp the motor's back-e.m.f., that C_1 limits unwanted RFI, and that R_1 limits Q_1's base current to about 52 mA with a 6 V input, and that under this condition Q_1 provides a maximum motor current of about 1 A.

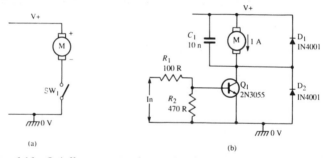

Figure 6.16 *On/off motor control using (a) electro-mechanical and (b) transistor switching*

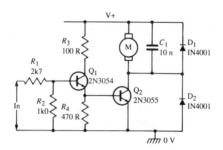

Figure 6.17 *Transistor motor switch with increased sensitivity*

In the above circuit, Q_1's 52 mA base current is provided via the external drive circuitry; if desired, the drive current can be reduced to a mere 2 mA or so by adding a buffer transistor as shown in *Figure 6.17*; note in this case that R_3 limits Q_1's base current to a safe value.

Direction control, using dual supplies

The rotational direction of a permanent magnet DC motor can be reversed by simply reversing the polarity of its supply connections. If the motor is powered via dual (split) supplies, this can be achieved via a single-pole switch connected as in *Figure 6.18*, or via transistor-aided switching by using the circuit of *Figure 6.19*.

In *Figure 6.19*, Q_1 and Q_3 are biased on and Q_2 and Q_4 are cut off (with Q_2's base-emitter junction reverse biased) when SW_1 is set to the *forward* position, and Q_2 and Q_4 are biased on and Q_1 and Q_3 are cut off (with Q_1's base-emitter junction reverse biased) when

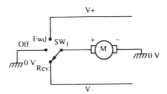

Figure 6.18   *Switched motor-direction control, using dual (split) power supplies*

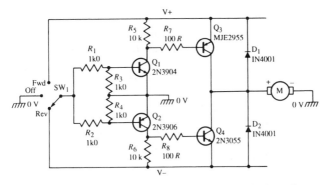

Figure 6.19   *Transistor-switched direction control, using dual supplies*

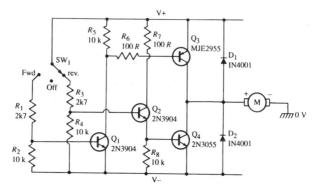

Figure 6.20   *Transistor-switched motor-direction control, using dual power supplies but a single-ended input*

SW_1 is set to the *reverse* position. Note that if this circuit is used with supply values greater than 12 V, diodes must be wired in series with the Q_1 and Q_2 base-emitter junctions, to protect them against breakdown when reverse biased.

The *Figure 6.19* circuit uses double-ended input switching, and this makes it difficult to replace SW_1 with electronic control circuitry in 'interfacing' applications. *Figure 6.20* shows how the design can be modified to give single-ended input switching control, making it easy to replace SW_1 with electronic switching. In this circuit, Q_1 and Q_3 are biased on and Q_2 and Q_4 are cut off when SW_1 is set to the forward position, and Q_2 and Q_4 are on and Q_1 and Q_3 are off when SW_1 is set to reverse.

Direction control, using single-ended supplies

If a permanent magnet DC motor is powered from single-ended supplies, its direction can be controlled via a double-pole switch connected as in *Figure 6.21*, or via a bridge-wired set of switching

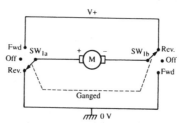

Figure 6.21   *Switched motor-direction control, using a single-ended power supply*

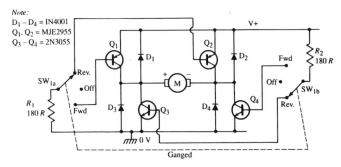

Figure 6.22 *Transistor-switched motor-direction control circuit, using single-ended supplies*

transistors connected in the basic form shown in *Figure 6.22*. In the latter case, Q_1 and Q_4 are turned on and Q_2 and Q_3 are off when SW_1 is set to the forward position, and Q_2 and Q_3 are on and Q_1 and Q_4 are off when SW_1 is set to reverse. Diodes D_1 to D_4 are used to protect the circuit against possible damage from motor back-e.m.f.s, etc.

Figure 6.23 shows how the above circuit can be modified to give

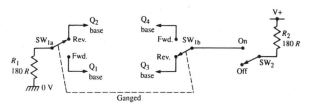

Figure 6.23 *Alternative switching for the Figure 6.22 circuit, using separate FWD/ REV (SW_1) and on/off (SW_2) switches*

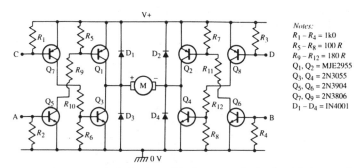

Figure 6.24 *Transistor-switched motor-direction control circuit with increased sensitivity*

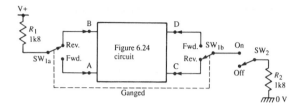

Figure 6.25 *Manual switching connections for use with Figure 6.24*

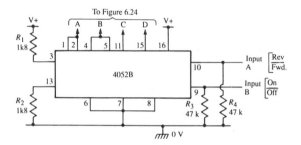

Figure 6.26 *Electronic control switching circuit for use with Figure 6.24*

Figure 6.26 circuit states						Figure 6.24 transistor states			
Inputs		Outputs				Q_1	Q_2	Q_3	Q_4
A|R/F|	B |on/off|	A	B	C	D				
0	0	1	X	X	X	On	X	X	X
0	1	1	X	X	O	On	X	X	On
1	0	X	1	X	X	X	On	X	X
1	1	X	1	0	X	X	On	On	X

X = Off or open circuit

Figure 6.27 *Truth table of the Figure 6.24 and Figure 6.26 circuits when they are interconnected*

alternative switching control via independent forward/reverse (SW_1) and on/off (SW_2) switches. A very important point to note about this configuration is that it causes Q_1 or Q_2 to be turned on at all times, with the on/off action being applied via Q_3 or Q_4, thus enabling the motor currents to collapse very rapidly (via the Q_1–D_2 or Q_2–D_1 loop) when the circuit is switched off. This so-called 'flywheel' action is vital if SW_2 is replaced by a pulse-width modulated (PWM) electronic switch, enabling the motor speed to be electronically controlled (this technique will be described shortly).

A weakness of the simple *Figure 6.22* circuit is that it uses fairly high base drive currents, which must be supplied via the switching circuitry. *Figure 6.24* shows a more sensitive version of the circuit, which required input control currents (to the A, B, C, D terminals) of only a few milliamperes.

The above circuit can be controlled manually via a pair of switches by using the connections shown in *Figure 6.25*, in which SW_1 controls the forward/reverse action and SW_2 controls the on/off action, or it can be controlled electronically by using the circuit of *Figure 6.26*, in which a 4052B CMOS IC is used as a ganged 2-pole 4-way bilateral switch (see *Figure 3.7*) that can be controlled via logic-0 or logic-1 signals applied to its A or B input pins, to give independent forward/reverse and on/off (or PWM speed control) actions.

Not that both of these circuits are configured to give the 'flywheel' type of switching action already described. *Figure 6.27* illustrates this point by showing the truth table that occurs when the *Figure 6.24* and *6.26* circuits are interconnected.

Motor speed control

The rotational speed of a DC motor is directly proportional to the mean value of its supply voltage; motor speed can thus be varied by altering either the value of its DC supply voltage, or, if the motor is operated in the switched-supply mode, by varying the m/s-ratio of its supply.

Figure 6.28 shows how variable-voltage speed control can be

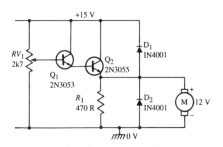

Figure 6.28 *Variable-voltage speed-control of a 12 V DC motor*

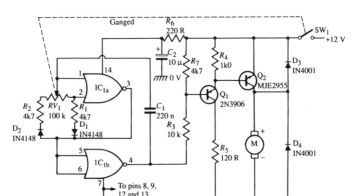

Figure 6.29   *Switched-mode speed-control of a 12 V DC motor*

obtained via variable pot RV_1 and compound emitter follower Q_1–Q_2, which enable the motor's DC voltage to be varied from zero to 12 V. This type of circuit gives fairly good speed control and self-regulation at medium to high speeds, but gives very poor low-speed control and slow-start operation. This type of circuit is thus used mainly in limited-range speed-control applications.

Figure 6.29 shows an example of a switched-mode speed-control circuit. Here, IC_1 acts as a 50 Hz astable multivibrator that generates a rectangular output with a m/s-ratio fully variable from 20:1 to 1:20 via RV_1, and this waveform is fed to the motor via Q_1 and Q_2. The motor's mean supply voltage (integrated over a 50 Hz period) is thus fully variable via RV_1, but is applied in the form of high-energy pulses with peak values of 12 V. This type of circuit thus gives excellent full-range speed control and generates high torque even at very low speeds; its degree of speed self-regulation is proportional to the mean value of applied voltage.

Model-train speed-controller

Figure 6.30 shows how the switched-mode principle can be used to make an excellent 12 V model-train speed-controller that enables speed to be varied smoothly from zero to maximum without jerkiness. The maximum available output current is 1.5 A, but the unit incorporates short-circuit sensing and protection circuitry that

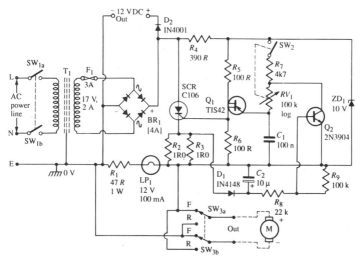

Figure 6.30 *Model train speed-controller circuit with automatic short-circuit protection*

automatically limits the output current to a mean value of only 100 mA if a short occurs on the track. The circuit operates as follows:

The circuit's power line voltage is stepped down via T_1 and full-wave (bridge) rectified via BR_1, to produce a raw (unsmoothed) DC supply that is fed to the model train (via the track rails) via the series-connected SCR and direction control switch SW_3. At the start of each raw DC half-cycle the SCR is off, so DC voltage is applied (via R_4 and ZD_1) to UJT Q_1 and its associated C_1–RV_1 (etc.) timing circuitry, and C_1 starts to charge up until eventually the UJT fires and triggers the SCR; as the SCR turns on it saturates, removing power from Q_1 (which thus resets) and feeding the rest of the power half-cycle to the model train via $R_2//R_3$ and SW_3. This timing/switching process repeats in each raw DC half-cycle (i.e., at twice the power line frequency), giving a classic phase-triggered power control action that enables the train speed to be varied over a wide range via RV_1.

Note that the circuit's output current passes through $R_2//R_3$, which generate a proportional output voltage that is peak-detected and stored via D_1–C_2 and fed to Q_2 base via R_9–R_9. The overall action is such that, because of the voltage storing action of C_2, Q_2 turns on and disables the UJT's timing network (thus preventing the SCR from firing) for several half-cycles if the peak output current exceeds 1.5 A. Thus, if a short occurs across the track the half-cycle output current is limited to a peak value of a few amperes by the circuit's internal resistance, but the protection circuitry ensures that the SCR fires only once in, say, every fifteen half-cycles, thus limiting the *mean* output current to only 100 mA or so.

An automatic track cleaner

Note that the raw DC output of the above circuit is available (via isolating diode D_2) via a pair of output terminals, and can be used to power model railway accessories such as automatic track cleaners, etc.

A major problem with model railways is that of maintaining electrical contact between the train pick-up wheels and the track, which both tend to pick up dirt and oxidization. This problem can be overcome by feeding the train control signals to the track via a load-sensing high-frequency low-power high-voltage generator or 'track cleaner', which harmlessly cuts its way through the existing dirt or oxidization. *Figure 6.31* shows an example of such a circuit, and its controller-to-track circuit connections.

The track cleaner is a modified blocking oscillator, tuned to operate at about 100 KHz by the inductance of step-up transformer T_1 (which is wound on a small ferrite core) and by the values of C_2 and C_4 (which minimizes the unwanted effects of track capacitance). The oscillator generates several hundred volts peak-to-peak

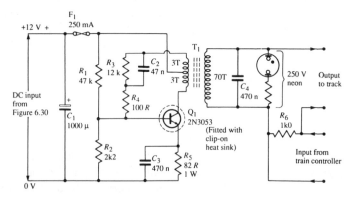

Figure 6.31   *Model railway track cleaner*

on T_1 secondary, but at a fairly high impedance (and thus harmless) level; oscillation ceases if the output is heavily loaded.

T_1's secondary is wound with fairly heavy guage wire, through which the train control signals are fed to the track. Thus, when electrical contact is made between a train motor and the track the resulting low impedance kills the oscillator, and only the train control signals reach the track, but if the contact is interrupted by dirt, etc., the resulting high impedance automatically enables the oscillator to work, and the resulting high-frequency high-voltage (plus train control) signals rapidly break through the interruption and re-establish electrical contact.

Note that a neon lamp (plus resistor) is wired across T_1 secondary, and illuminates when the track cleaner is active, thus indicating loss of track contact. R_6 ensures that only a very small part of the oscillator output voltage can be fed to the train controller terminals when the cleaner is active.

Motor speed regulation

Motor speed regulators are meant to keep motor speed fairly constant in spite of wide variations in the control circuit's supply voltage and in the motor loading conditions, etc. *Figure 6.32* shows an example of a regulator circuit that is designed to simply keep the motor's applied voltage constant in spite of wide variations in supply voltage and temperature.

This circuit is designed around a 317K 3-terminal variable voltage regulator IC which (when fitted to a suitable heat sink) can supply output currents up to 1.5 A and has an output that is fully protected against short circuit and overload conditions. With the component values shown, the output is fully variable from 1.25 V to

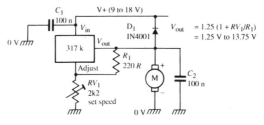

Figure 6.32 *Simple motor-speed controller/regulator*

13.75 V via RV_1, provided that the supply voltage is at least three volts greater than the desired output value.

Figure 6.33 shows a popular type of regulator circuit that is widely used (with minor variations) in cassette recorders, and compensates for variations in both battery voltage and motor loading conditions. The motor current is controlled via series transistor Q_1 and is monitored via R_2, and Q_1 is controlled via Q_2. The diagram shows the circuit voltages obtained when the motor is operating at 6 V and is drawing 100 mA. Note that Q_2's emitter is biased 1.2 V below the motor value via D_1–D_2–R_3, and Q_1's base is set at a fraction of the Q_1-collector value via R_4–RV_1–R_5; all changes in motor or Q_1 collector voltages thus effect Q_2's emitter and base bias values, but the changes are always least on the base.

Thus, any drop in the circuit's supply voltage tends to decrease the motor's voltage and make Q_2's emitter fall further below its base value, thus driving Q_2 and Q_1 harder on and self-compensating for the supply reduction. Similarly, any increase in motor loading tends to slow the motor and increase the motor current and the R_2 volt-drop, thus raising the relative value of Q_1's collector voltage and making Q_2's base voltage rise further above that of the emitter, thus driving Q_2 and Q_1 harder on and increasing the motor drive, to self-compensate for the increased motor loading. This simple circuit thus gives automatic regulation of motor speed to compensate for variations in supply voltage or loading conditions; some thermal compensation is also given via D_1 and D_2. The motor speed can be varied over a limited range via RV_1.

Finally, *Figure 6.34* shows a high-performance regulator circuit

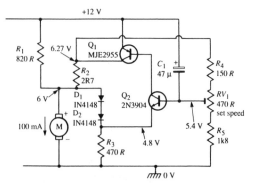

Figure 6.33   *Popular motor-speed regulator circuit*

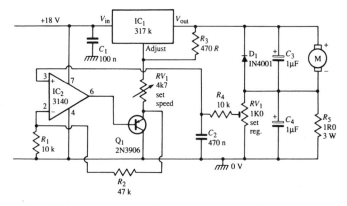

Figure 6.34 *High-performance variable-speed regulator circuit*

that can be used in wide-range variable-speed applications, such as controlling 12 V DC minidrills, etc. Here, the motor is again powered via the output of a 317K 3-terminal variable voltage regulator IC, but in this case the motor current is monitored via R_5–RV_2, which feed a proportional voltage to the input of the IC_2–Q_1 non-inverting DC amplifier, to generate a Q_1 emitter voltage directly proportional to the motor's load current.

Now, the output voltage of this circuit equals the normal output value of the 317K IC (which is variable from 1.25 to 13.75 V via RV_1) plus the voltage on Q_1's emitter; consequently, any increase in motor loading makes the circuit's output voltage rise, to automatically increase the motor drive and hold its speed reasonably constant. To initially set up the circuit, simply set the motor speed to about one-third of maximum via RV_1, then lightly load the motor and set RV_2 so that the speed remains similar in both loaded and unloaded states.

Two-phase motor driver

Two-phase (AC) motors are synchronous machines, and low-voltage versions are sometimes used as precision phonograph turntable drivers. *Figure 6.35* shows a circuit that can drive 8 Ω motor windings at up to 3 W each, at frequencies between 45 Hz and 65 Hz. The circuit is designed around an LM377 dual 3 W audio

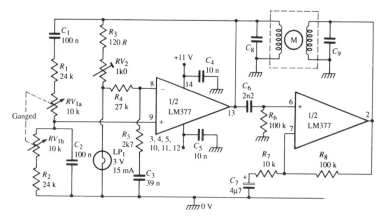

Figure 6.35 *Two-phase motor drive*

power amplifier IC, driven from a split supply, and operates as follows.

The IC's left-hand half is wired as a Wien bridge oscillator, with frequency variable between 45 Hz and 65 Hz via RV_1 and with amplitude stabilized via RV_2 and filament lamp LP_1; the output is fed directly to one motor phase winding, and to the other via the IC's right-hand half, which acts as an 85° phase shifter (C_6–R_6 shifts the phase by 85° but attenuates by a factor of ten at 60 Hz, and the IC half gives unity phase shift and a gain of ten at 60 Hz). Circuit stability is assured via decoupling networks C_3–R_4–R_5, C_4, and C_5, and the motor windings are tuned to a mid-frequency value via C_8 and C_9.

Servomotor systems

A servomotor is a conventional electric motor with its output coupled (usually via a speed-reduction gearbox) to a movement-to-data translator, such as a potentiometer or a tachogenerator. *Figure 6.36* shows a controller that can be used to give proportional movement (set via RV_2) of a servomotor with a pot (RV_1) output; the motor can be any 12 to 24 V type that draws less than 700 mA of current. Here, RV_1 and RV_2 are wired as a Wheatstone bridge, and the IC (a dual 4 W power amplifier) is wired as a bridge-configured motor-driving difference amplifier. The circuit action is such that any movement of RV_2 upsets the bridge balance and generates a

RV_1–RV_2 difference voltage that is amplified and fed to the motor, making its shaft rotate and move RV_1 to restore the bridge balance; RV_1 thus 'tracks' the movement of RV_2, which can thus be used to remote-control the shaft position.

Figure 6.37 shows, in block diagram form, how a tachogenerator type of servomotor system can be used to give precision speed control of a phonograph turntable. Here, the motor drives the turntable via a conventional belt drive mechanism, and the turntable edge is patterned with equally-spaced reflective strips that are monitored by an optoelectronic tachogenerator which produces an output signal proportional to the turntable speed. The phase and frequency of this signal is compared with that of a precision oscillator, to give an output that is used to control the motor drive circuit, thus holding the turntable at the precise speed selected by the operator. Several manufacturers produce dedicated ICs for use in this type of application.

One of the best known types of servomotor is that used in digital proportional remote control systems; these devices actually consist of a special IC plus a motor and a reduction gearbox that drives a pot and gives a mechanical output. *Figure 6.38* shows the block diagram of one of these systems, which is driven via a variable-width (1 mS to 2 mS) input pulse that is repeated once every 15 mS or so (the frame time). The input pulse width controls the position of the servo's mechanical output; at 1 mS the servo output may, for example, be full left, at 1.5 mS neutral, and at 2 mS full right. The system operates as follows.

Each input pulse triggers a 1.5 mS 'deadband' pulse generator and a variable-width pulse generator controlled (via RV_1) by the gearbox output; these three pulses are fed to a width comparator

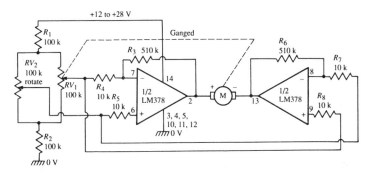

Figure 6.36 *Proportional-movement (servomotor) controller*

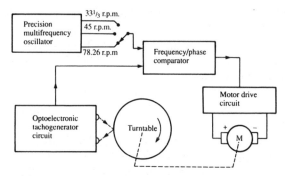

Figure 6.37 *Block diagram showing a tachogenerator servomotor used to give precision speed control of a phonograph turntable*

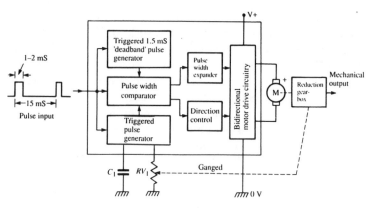

Figure 6.38 *Block diagram showing basic digital proportional-control servomotor system*

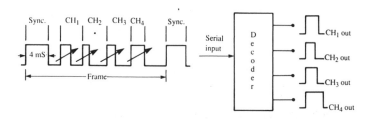

Figure 6.39 *Waveforms of a 4-channel digital proportional-control system*

that gives one output that gives direction control of the motor drive circuitry, and another that (when fed through a pulse-width expander) controls the motor speed, thus making the servomotor's RV_1-driving mechanical output rapidly follow any variations in the width of the input pulse.

Servomotors of the above type are usually used in multichannel remote control systems, as shown in the basic 4-channel system of *Figure 6.39*. Here, a serial data input is fed (via some form of data link) to the input of a suitable decoder. Each input frame comprises a 4 mS synchronization pulse followed by four variable-width (1 mS to 2 mS) sequential 'channel' pulses. The decoder simply converts the four channel pulses to parallel form, enabling each pulse to be used to control a servomotor.

Digital servomotor circuits

Digital proportional servomotor units are widely available in both kit and ready-built forms, and are usually designed around either the Ferranti ZN409CE or the Signetics NE544N servo amplifier ICs. *Figures 6.40* and *6.41* show practical application circuits for

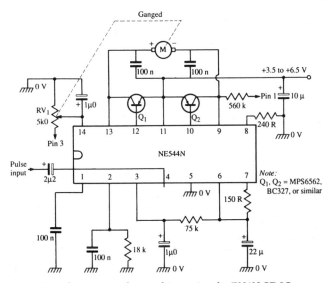

Figure 6.40 *Digital proportional servo driver using the ZN409CE IC*

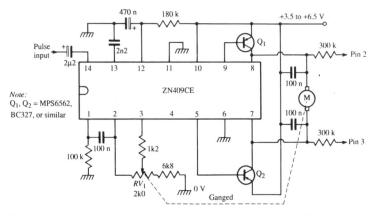

Figure 6.41   *Digital proportional servo driver using the NE544N IC*

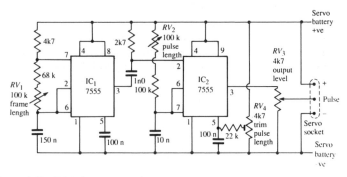

Figure 6.42   *Digital proportional servo tester circuit*

both of these IC types, with component values suitable for input pulse lengths in the 1 to 2 mS range and frame length of about 18 mS nominal.

Finally, to complete this chapter, *Figure 6.42* shows a general-purpose tester for use with the above types of servo. This unit is powered from the servo's supply battery (nominally about 5 V) and simply feeds normal input pulses to the servo via a standard servo socket. The frame length is variable from 13 mS to 28 mS via RV_1, and the pulse length is variable from 1 mS to 2 mS via RV_2 and can be trimmed to give a precise 1.5 mS mid-scale value via RV_4; the output pulse level is variable via RV_3.

The circuit is designed around two 7555 ICs, which are CMOS versions of the 555 timer chip and give stable operation at supply values down to 3 V. IC_1 is configured as a free running astable multivibrator, and generates the frame times, and its output triggers IC_2, which is configured as a monostable multivibrator and generates the output test pulses.

7 Audio power control circuits

Audio power control circuits can, as far as this chapter is concerned, be defined as those that convert an audio input signal into an accurately reproduced acoustic output (via a load speaker), generating a minimum of distortion in the process. A great many circuits of this type, together with audio preamplifier and filter circuits, etc., have already been described in my book entitled *Audio IC Circuits Manual*, which presents more than 240 audio circuits and diagrams, so in this chapter we will simply look at a small selection of practical IC-based single-channel (mono) audio power control circuits.

Low-power circuits

Simple audio amplifiers with output powers up to a few hundred milliwatts can be easily and cheaply built using little more than a standard op-amp and a couple of general-purpose transistors. The popular 741 op-amp, for example, can supply peak output currents of at least 10 mA and peak output voltage swings of at least 10 V onto a 1k0 load when powered from a dual 15 V supply, thus giving peak outputs of 100 mW under these conditions.

Figure 7.1 shows how the 741 op-amp can be used as a low-level audio power amplifier, using a dual power supply. The external load is direct-coupled between the op-amp output and ground, and the two input terminals and ground-referenced. The op-amp is used in the non-inverting mode, and has a voltage gain of $\times 10$ ($= R_1/R_2$) and input impedance of 47k ($= R_3$).

Figure 7.2 shows how to use the 741 with a single-ended power

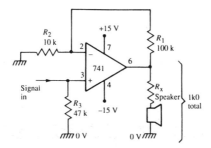

Figure 7.1 *Low-power amplifier using dual power supplies*

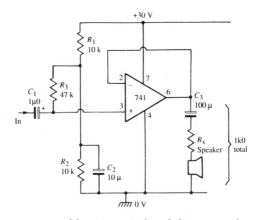

Figure 7.2 *Low-power amplifier using a single-ended power supply*

supply. Note in this case that the load is a.c.-coupled between the output and ground, and that the output is biased to a quiescent half-supply voltage value (to give maximum output voltage swing) via the R_1–R_2 potential divider. The op-amp is shown used in the unity-gain non-inverting mode, with an input impedance of 47k $(= R_3)$.

Note in the above two circuits that the external load must have an impedance of at least 1k0. If the external speaker has an impedance less than this, resistor Rx can be connected as shown to raise the impedance to the 1k0 value, but inevitably reduces the amount of power reaching the actual speaker.

In practice, the available output current (and thus power) of an op-amp can easily be boosted via a transistor complementary emitter follower network wired into its output. *Figure 7.3* shows such a circuit, using a single-ended supply. Here, the Q_1–Q_2

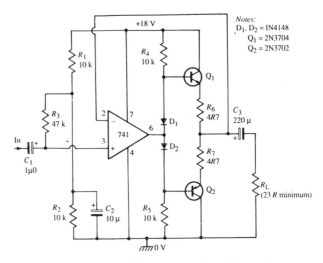

Figure 7.3   *Op-amp power amplifier using a single-ended supply*

emitter follower has slight forward bias applied via D_1 and D_2 and is wired into the circuit's negative feedback loop to minimize crossover distortion, etc. The circuit gives unity overall voltage gain and can supply output currents up to 350 mA peak or 50 mA rms into a load of 23 Ω minimum, i.e., it can provide powers up to 280 mW into such a load.

Simple audio PA ICs

If output powers greater than a couple of hundred milliwatts are needed, the easiest solution is to use a dedicated audio power amplifier (PA) IC. A large range of such ICs is readily available, in both mono (single amplifier) and dual (two amplifiers) forms, and with maximum output power ratings ranging from about 325 mW to 22 W. *Figures 7.4* to *7.7* show practical application circuits using mono ICs with power ratings in the 325 mW to 5 W range.

The *Figure 7.4* circuit is based on the LM386, which is housed in an 8-pin DIL package and has an output power rating of 325 mW; it can use DC supplies in the 4 V to 15 V range, consumes a quiescent current of 4 mA, and is useful in many battery-powered applications. The diagram shows the IC used in the non-inverting mode; the voltage gain can be set via pins 1 and 8, and equals ×20 when

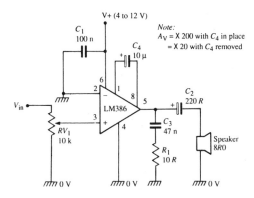

Figure 7.4 *An LM386 (325 mW) amplifier*

these pins are open-circuit, or ×200 when they are a.c.-shorted via C_4. Note that C_1 is used to RF-decouple the positive supply pin (pin-6), and R_1–C_3 is an optional Zobel network that ensures high frequency (HF) stability when feeding an inductive speaker load.

The *Figure 7.5* circuit is based on the LM390, which is designed to operate from 4 V to 9 V supplies and can feed 1 W into a 4R0 load when using a 6 V supply. The IC is housed in a 14-pin DIL package with an internal heat sink connected to pins 3–4–5 and 10–11–12. The diagram shows the IC used in the non-inverting mode; the voltage gain can be set via pins 2 and 6, and equals ×20 when these pins are open-circuit, or ×200 when they are a.c.-shorted via C_5. R_3–C_4 is an optional Zobel network.

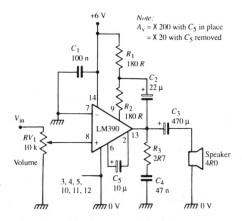

Figure 7.5 *An LM390 (1 W) amplifier*

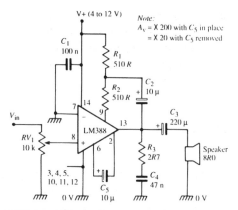

Figure 7.6   *An LM388 (1.5 W) amplifier*

The *Figure 7.6* circuit is based on the LM388, which can be regarded as a modified version of the LM386 fitted into a 14-pin DIL package with integral heat sink. The IC can feed 1.5 W into an 8R0 speaker when powered from a 12 V supply. The circuit's voltage gain can be set via pins 2 and 6, and equals ×20 when these pins are open-circuit, or ×200 when they are a.c.-shorted via C_5.

Finally, the *Figure 7.7* circuit can use either the LM380 or the LM384 IC, and can supply output powers up to a maximum of 5 W. The LM380 is probably the best known of all audio PA ICs; it can use any supply in the 8 V to 22 V range, and can deliver 2 W into an 8R0 load when operated with an 18 V supply (but needs a good external heat sink to cope with this power level). It has ground-referenced differential input terminals, a fully protected short-circuit proof output, and gives a fixed voltage gain of ×50 (34 dB).

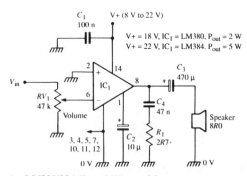

Figure 7.7   *An LM380/384 (2 or 5 W) amplifier*

The LM384 is simply an uprated version of the LM380, capable of operating at supply values up to 26 V and delivered 5.5 W into an external load. Both types of IC are housed in 14-pin DIL packages in which pins 3–4–5 and 10–11–12 are meant to be thermally coupled to an external heat sink.

Bridge-mode operation

When an IC power amplifier is used in the single-ended output mode shown in *Figure 7.4* to 7.7, the peak available output power equals V^2/R, where V is the peak available output voltage and R is the output load impedance. Note, however, that available output power can be increased by a factor of four by connecting a pair of amplifier ICs in the 'bridge' configuration shown in *Figure 7.8*, in which the peak available load power equals $(2V)^2/R$. This power increase can be explained as follows.

In an ordinary single-ended amplifier, one end of the output load is usually ground; in *Figure 7.8*, however, both ends of load R_L are floating and driven in antiphase, and the voltage across R_L equals the difference between the A and B values. The diagram shows the waveforms applied to the load via a 10 V peak-to-peak square-wave

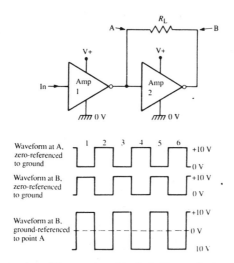

Figure 7.8 *A pair of amplifiers connected in the bridge mode give a peak output of $(2V)^2/R$ W, i.e., four times the power of a single-ended circuit*

input signal. Note that although waveforms A and B each have peak values of 10 V relative to ground, the two signals are in antiphase. Thus, during period 1 point B is 10 V positive to A and is thus seen as being at +10 V, but in period 2 point B is 10 V negative to point A and is thus seen as being at −10 V. Consequently, if point A is regarded as a zero voltage reference point, it can be seen that point B varies from +10 V to −10 V between periods 1 and 2, giving a total voltage change of 20 V across R_L. Similar changes occur in subsequent waveform periods.

Thus, the load in a 10 V bridge-driven circuit sees a total voltage swing of 20 V peak-to-peak, or twice the single-ended voltage value. Since doubling the drive voltage results in a doubling of drive current, and power is equal to the *V–I* product, the bridge-driven circuit thus produces four times more power output than a single-ended one.

Figure 7.9 and *7.10* show a couple of practical examples of bridge-driven audio power amplifiers. The *Figure 7.9* circuit uses a pair of LM390 ICs and can feed 2.5 W into a 4R0 load when using a 6 V supply. The *Figure 7.10* circuit can use a pair LM380 or LM384 ICs, and can deliver up to 10 W into an 8R0 load. Note in each of these circuits that the speaker is direct coupled to the outputs of the power amplifier ICs, eliminating the need for bulky and expensive electrolyte decoupling capacitors.

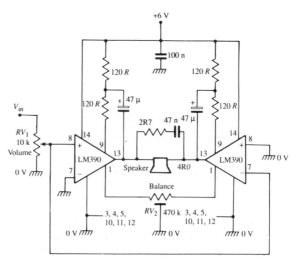

Figure 7.9   *LM390 bridge amplifier delivers 2.5 W into a 4R0 load*

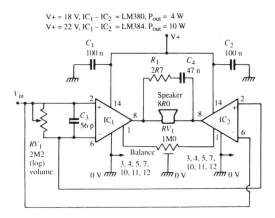

Figure 7.10 *4 or 10 W bridge-configured amplifier*

Dual IC applications

Audio power amplifier ICs are widely available in both mono and dual forms. Dual types are very handy for making stereo or bridge-driven amplifiers, and the LM377 dual 2 W, the LM378 dual 4 W, and the LM379 dual 6 W are among the best known of these types. These three devices differ mainly in their voltage/power ratings and packaging styles. The LM377 and LM378 are each housed in a 14-pin DIL package with pins 3–4–5 and 10–11–12 connected to an internal heat sink, and the LM379 is housed in a 14-pin package with an attached heat sink.

The LM377/378/379 range of ICs is very easy to use: the input stages of each half are meant to be d.c.-biased to half-supply volts, and a bias generator is built into each IC for this purpose. *Figure 7.11* shows how to wire each IC as a simple stereo amplifier powered from a single-ended supply. Each amplifier is wired in the inverting mode, is biased by connecting its non-inverting input pin to the IC's bias terminal (pin-1 on the LM377 or LM378, or pin-14 on the LM379), and has its closed-loop voltage gain set at ×50 by the ratio of R_2/R_1 or R_4/R_3. The table shows the typical performance of this circuit.

If desired, the output power of each IC half can be boosted to a higher value via a couple of external power transistors; *Figure 7.12* shows how the available output power of one half of the LM378 can be boosted to 15 W in this way. This remarkably simple circuit

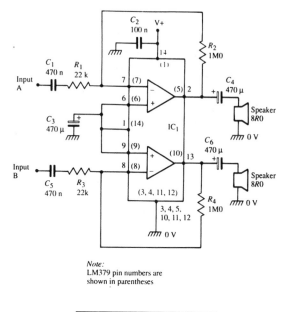

Note:
LM379 pin numbers are
shown in parentheses

	IC$_1$		
	LM377	LM378	LM379
V+ (max.)	18 V	24 V	28 V
P_{out}/CH	2 W	3 W	4 W
e_{in} (max.)	80 mV	100 mV	115 mV
A_V (approx.)	50	50	50
Z_{in}	22 k	22 k	22 k

Typical performance of the
inverting stereo amplifier

Figure 7.11 *Simple stereo amplifier using the LM377, LM378, or LM379 dual amplifier ICs*

generates a typical THD of only 0.05 per cent at 10 W output; at very low power levels Q_1 and Q_2 are inoperative and power is fed to the speaker via R_2; at higher power levels Q_1 and Q_2 act as a normal complementary emitter follower and provide most of the power drive to the speaker. R_2 and the base-emitter junctions of Q_1–Q_2 are effectively wired into the circuit's negative feedback loop, thus minimizing signal crossover distortion.

Finally, *Figure 7.13* shows how the two halves of a LM377, LM378, or LM379 can be used to make a bridge-configured mono amplifier which can feed relatively high power levels to a direct-coupled speaker load.

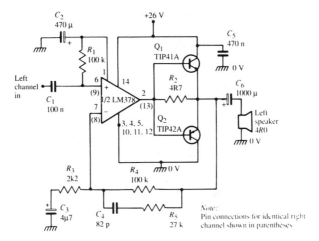

Figure 7.12 *One channel of a 15 W per channel stereo amplifier using a single-ended supply*

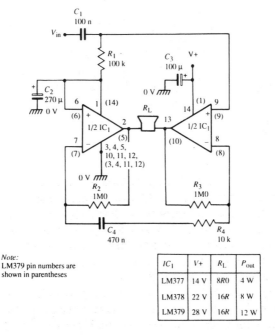

Note:
LM379 pin numbers are
shown in parentheses

IC_1	V+	R_L	P_{out}
LM377	14 V	8R0	4 W
LM378	22 V	16R	8 W
LM379	28 V	16R	12 W

Figure 7.13 *Bridge amplifier using dual ICs*

ICs for cars

Several manufacturers produce audio power amplifier ICs for specific use in in-car entertainment systems. Two of the best known devices of this type are the identical LM383 and TDA2003 8 W devices, which are each housed in a 5-pin TO–220 plastic package with integral heat sink. At a car's normal 'running' supply voltage of 14.4 V each IC can deliver 5.5 W into a 4R0 load or 8.6 W into a 2R0 load; each IC can in fact operate with any 5 V to 20 V supply voltage and can supply peak output currents of 3.5 A.

The LM383 (TDA2003) is an easy device to use. *Figure 7.14* shows it wired as a 5.5 W in-car amplifier, with the IC wired in the non-inverting mode with its closed-loop voltage gain set at ×100 via R_1–R_2. C_2 and C_4 ensure the IC's HF stability; C_4 must be direct-wired between pins 3 and 4.

If desired, a pair of LM383 or TDA2003 ICs can be wired in the bridge configuration, to act as a 16 W in-car mono amplifier. *Figure 7.15* shows the circuit connections; note that preset pot RV_1 is used to balance the quiescent output voltages of the two ICs, to minimize the circuit's quiescent operating current.

Some of the available in-car audio PA ICs are dual types, which can be used in either the stereo or the bridge-configured modes. One of the most popular types of 'dual' IC is the TDA2005M, which has its two halves internally wired in the bridge mode, to provide up to 20 W of drive into a 2R0 load from the vehicle's 14.4 V (nominal) supply. The IC is housed in an 11-pin package, as

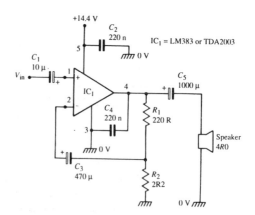

Figure 7.14   *LM383 (TDA2003) 5.5 W in-car amplifier*

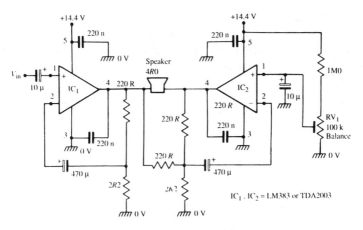

Figure 7.15 *LM383 (TDA2003) 16 W in-car amplifier*

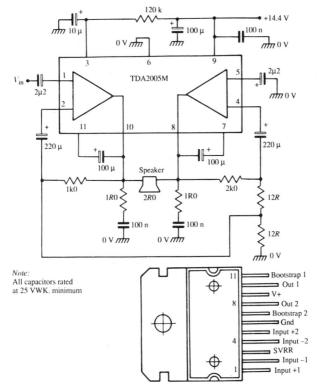

Figure 7.16 *TDA2005M 20 W in-car power booster*

shown in *Figure 7.16*, which also shows a practical TDA2005M application circuit. Note that all of this circuit's capacitors must be rated at 25 V minimum.

Hi-fi amplifiers

Several manufacturers produce high-performance mono audio power amplifier ICs that are suitable for direct use in hi-fi systems with per-channel output power ratings in the range 8 to 22 W. To conclude this chapter, *Figures 7.17* to *7.20* show four practical application circuits for ICs of this type.

The *Figure 7.17* circuit has an output power rating of 8 W, and is designed around a TDA2006 IC. This popular high-quality device typically generates less than 0.1 per cent distortion when feeding 8 W into a 4R0 speaker, and is housed in a 5-pin TO–220 package with an electrically insulated heat tab which can be bolted directly to an external heat sink without need of an insulation washer. In the diagram, the IC's non-inverting input pin is biased at half-supply volts via R_3 and the R_1–R_2 potential divider, and the voltage gain is set at ×22 via R_5/R_4. D_1 and D_2 protect the IC's output against damage from the speaker's back e.m.f.s, and R_6–C_6 form a Zobel network.

The *Figure 7.18* circuit is almost identical to the above, but has an output power rating of 15 W and is designed around a TDA2030 IC.

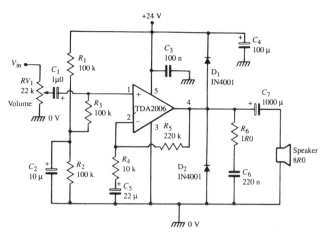

Figure 7.17   *TDA2006 8 W hi-fi amplifier*

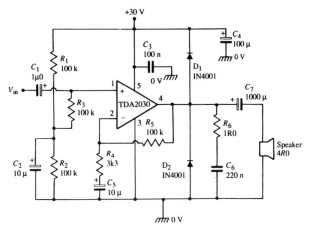

Figure 7.18 *TDA2030 15 W hi-fi amplifier*

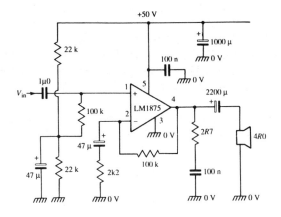

Figure 7.19 *LM1875 20 W hi-fi amplifier*

This very popular IC can be regarded as an up-rated version of the TDA2006, and is housed in the same 5-pin TO–220 package with insulated heat tab. It can operate with single-ended supplies of up to 36 V; when used with a 28 V supply it gives a guaranteed output of 12 W into 4R0, or 8 W into 8R0. Typical THD is 0.05 per cent at 1 kHz at 7 W.

The *Figure 7.19* circuit has an output power rating of 20 W into a 4R0 load, and is designed around an LM1875 IC. This high-quality device generates a mere 0.015 per cent THD at 20 W output, can

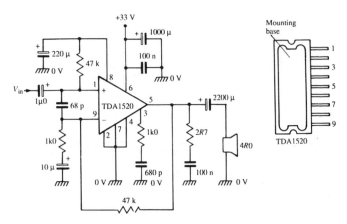

Figure 7.20 *TDA1520 22 W hi-fi amplifier*

operate at supply voltages as high as 60 V, and is housed in a 5-pin TO–220 package with an insulated heat tab.

Finally, the *Figure 7.20* circuit has an output power rating of 22 W into a 4R0 load, and is designed around a TDA1520. This is a very high performance IC which typically generates a mere 0.01 per cent distortion at 16 W output into a 4R0 load. The IC is housed in a 9-pin package (shown in the diagram) that can be bolted directly to an external heat sink without the need for insulating washers in single-ended supply applications.

8 DC power supply circuits

This final chapter deals with DC power supply systems and circuits, and is roughly divided into three sections. The first section deals with ways of deriving DC power from AC power lines, the next deals with DC voltage regulator circuits, and the final section deals with low-power voltage converter circuits which can, for example, be used to generate a higher-value or reversed-polarity voltage supply from an existing DC power source.

AC/DC converter basics

There are two basic ways of deriving a stable DC supply from an AC power line, and these are shown in basic form in *Figures 8.1* and *8.2*. The conventional way (*Figure 8.1*) is to use a step-down transformer and a rectifier and storage capacitor to generate (with an overall efficiency of about 85 per cent) an unregulated DC supply that is electrically insulated from the AC supply, and to then stabilize the DC output voltage via a linear regulator circuit, which

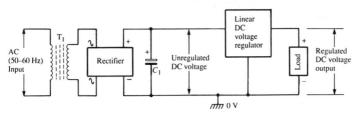

Figure 8.1 *Conventional regulated DC power supply: typical conversion efficiency is 45 per cent*

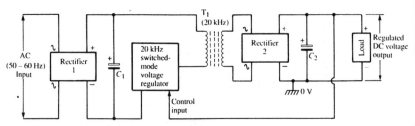

Figure 8.2 *Switched-mode regulated DC power supply; typical conversion efficiency is 80 per cent*

typically has an efficiency slightly better than 50 per cent, thus giving an overall power conversion efficiency of about 45 per cent.

Major advantages of these conventional systems are that they are simple and reliable, generate zero RFI, and give excellent ripple reduction and voltage regulation. The major disadvantage is that they are rather bulky and heavy, because (a) the transformer operates at the 50–60 Hz supply line frequency, and (b) a heat sink is needed to dissipate the power wasted by the low conversion efficiency. Systems of this type are used in virtually all radio and audio equipment, and in all RFI-sensitive instruments.

The alternative AC/DC conversion system uses the switched-mode technique shown in *Figure 8.2*. Here, the AC power line voltage is directly converted to DC via a rectifier and storage capacitor, and this DC is used to power a series-connected 20 kHz switched-mode (variable m/s-ratio) voltage regulator and isolating transformer, which has its output converted back to DC via another rectifier and storage capacitor; part of the DC output is fed back to the switched-mode voltage regulator, to complete a control loop.

Major advantages of the switched-mode system are that it is very efficient (typically about 80 per cent) and compact (because its heat/sink requirements are minimal and its isolating transformer works at 20 kHz). Major disadvantages are that it generates substantial RFI, and gives poorer ripple rejection and voltage regulation than the conventional system. Switched-mode systems are widely used in desktop calculators and computers and in other very compact equipment that is not RFI-sensitive.

Basic power supply circuits

Most modern electronic equipment uses the conventional type of power supply, and this is the only type dealt with in the rest of this

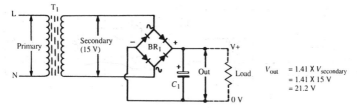

Figure 8.3 *Basic single-ended power supply using a single-ended transformer and bridge rectifier*

chapter. The basic element of this supply is the AC/DC converter, which consists of a transformer that converts the AC line voltage into an electrically insulated and more useful AC value, and a rectifier-filter combination that converts this into smooth DC of the desired voltage value.

Figures 8.3 to *8.6* show the four most widely used basic power supply circuits. The *Figure 8.3* design provides a single-ended DC

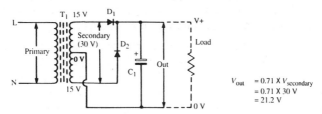

Figure 8.4 *Basic single-ended power supply using a centre-tapped transformer and two rectifiers*

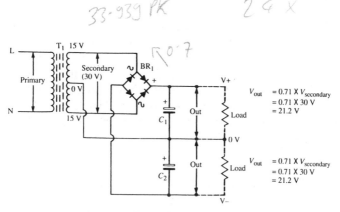

Figure 8.5 *Basic split or dual power supply using a centre-tapped transformer and bridge rectifier*

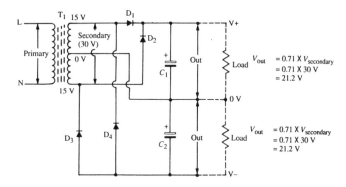

Figure 8.6 *Basic split or dual power supply using a centre-tapped transformer and individual rectifiers*

supply from a single-ended transformer and bridge rectifier combination, and gives a performance virtually identical to that of the *Figure 8.4* centre-tapped transformer circuit. The *Figures 8.5* and *8.6* circuits each provide split or dual DC supplies with nearly identical performances. The rules for designing these four circuits are quite simple, as follows.

Transformer-rectifier selection

The three most important parameters of a transformer are its secondary voltage, its power rating, and its regulation factor. The secondary voltage is always quoted in r.m.s. terms at full rated power load, and the power load is quoted in terms of voltamperes or watts. Thus, a 15 V 20 VA transformer gives a secondary voltage of 15 V r.m.s. when its output is loaded by 20 W. When the load is removed (reduced to zero) the secondary voltage rises by an amount implied by the *regulation factor*. Thus, the output of a 15 V transformer with a 10 per cent regulation factor (a typical value) rises to 16.5 V when the output is unloaded.

Note that the transformer's r.m.s. output voltage is *not* the same as the DC output voltage of the complete full-wave rectified power supply which, as shown in *Figure 8.7*, is in fact 1.41 times greater than that of a single-ended transformer, or 0.71 times that of a centre-tapped transformer (ignoring rectifier losses). Thus, a single-ended 15 V r.m.s. transformer with 10 per cent regulation gives an output of about 21 V at full rated load (just under 1 A at 20

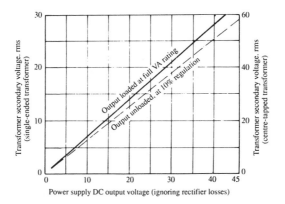

Figure 8.7 *Transformer selection chart. To use, decide on the required loaded DC output voltage (say, 21 V), then read across to find the corresponding transformer secondary voltage (15 V single-ended or 30 V centre-tapped)*

VA rating) and 23.1 volts at zero load. When rectifier losses are taken into account the output voltages are slightly lower than shown in the graph. In the two-rectifier circuits of *Figures 8.4* and *8.6* the losses are about 600 mV, and in the bridge circuits of *Figures 8.3* and *8.5* they are about 1.2 V. For maximum safety, the rectifiers should have current ratings at least equal to the DC output currents.

Thus, to select a transformer for a particular task, first decide the DC output voltage and current that is needed, to establish the transformer's minimum voltampere rating, then simply consult the graph of *Figure 8.7* to find the transformer secondary r.m.s. voltage that corresponds to the required DC voltage.

The filter capacitor

The purpose of the filter capacitor is to convert the rectifier's output into a smooth DC voltage; its two most important parameters are its working voltage, which must be greater than the off-load output value of the power supply, and its capacitance value, which determines the amount of ripple that will appear on the DC output when current is drawn from the circuit.

As a rule of thumb, in a full-wave rectified power supply operating from a 50–60 Hz power line, an output load current of 100 mA will cause a ripple waveform of about 700 mV peak-to-

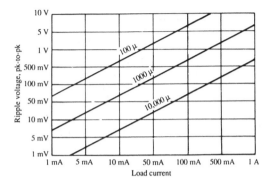

Figure 8.8 *Filter capacitor selection chart, relating capacitor size to ripple voltage and load current in a full-wave rectified 50–60 Hz powered circuit*

peak to be developed on a 1000 μF filter capacitor, the amount of ripple being directly proportional to the load current and inversely proportional to the capacitance value, as shown in the design guide of *Figure 8.8*. In most practical applications, the ripple should be kept below 1.5 V peak-to-peak under full load conditions. If very low ripple is needed, the basic power supply can be used to feed a 3-terminal voltage regulator IC, which can easily reduce the ripple by a factor of 60 dB or so at low cost.

Zener-based voltage regulator

Practical voltage regulators vary from simple Zener diode circuits designed to provide load currents up to only a few milliamperes, to fixed- or variable-voltage high-current circuits designed around dedicated 3-terminal voltage regulator ICs. Circuits of all these types are shown in the next few sections of this chapter.

Figure 8.9 shows how a Zener diode can be used to generate a

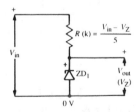

Figure 8.9 *This basic Zener 'reference' circuit is biased at about 5 mA*

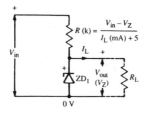

Figure 8.10 *This basic Zener 'regulator' circuit can supply load currents of a few tens of milliamperes*

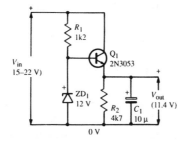

Figure 8.11 *This series-pass Zener based regulator circuit gives an output of 11.4 V and can supply load currents up to about 100 mA*

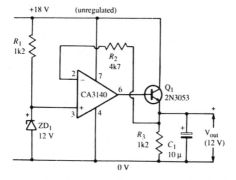

Figure 8.12 *This op-amp based regulator gives an output of 12 V at load currents up to 100 mA and gives excellent regulation*

fixed reference voltage by passing a current of about 5 mA through it from the supply line via limiting resistor R. In practice, the output reference voltage is not greatly influenced by sensible variations in the diode current value, and these may be caused by variations in the values of R or the supply voltage, or by drawing current from the output of the circuit. Consequently, this basic circuit can be made to function as a simple voltage regulator, generating output load currents up to a few tens of milliamperes by merely selecting the R value as shown in *Figure 8.10*.

Here, the value of R is selected so that it passes the maximum desired output current plus 5 mA; consequently, when the specified maximum output load current is being drawn the Zener passes only 5 mA, but when zero load current is being drawn it passes all of the R current, and the Zener dissipates maximum power; the power rating of the Zener must not be exceeded under this 'no load' condition.

The available output current of a Zener regulator can easily be increased by wiring a current-boosting voltage follower into its output, as shown in the series-pass voltage regulator circuits of *Figure 8.11* and *8.12*. In *Figure 8.11*, Q_1 acts as the voltage following current booster, and gives an output that is about 600 mV below the Zener value; this circuit gives reasonably good regulation. In *Figure 8.12*, Q_1 and the CA3140 op-amp form a precision current-boosting voltage follower that gives an output equal to the Zener value under all load conditions; this circuit gives excellent voltage regulation. Note that the output load current of each of these circuits is limited to about 100 mA by the power rating of Q_1; higher currents can be obtained by replacing Q_1 with a power Darlington transistor.

Fixed 3-terminal regulator circuits

Fixed-voltage regulator design has been greatly simplified in recent years by the introduction of 3-terminal regulator ICs such as the 78xxx series of positive regulators and the 79xxx series of negative regulators, which incorporate features such as built-in fold-back current limiting and thermal protection, etc. These ICs are available with a variety of current and output voltage ratings, as indicated by the 'xxx' suffix; current ratings are indicated by the first part of the suffix (L = 100 mA, blank = 1 A, S = 2 A), and the

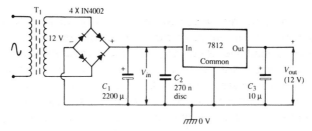

Figure 8.13 *Connections for using a 3-terminal positive regulator, in this case a 12 V 1 A '78' type*

voltage ratings by the last two parts of the suffix (standard values are 5 V, 12 V, 15 V and 24 V). Thus, a 7805 device gives a 5 V positive output at a 1 A rating, and a 79L15 device gives a 15 V negative output at a 100 mA rating.

3-terminal regulators are very easy to use, as shown in *Figures 8.13* to *8.15*, which show positive, negative and dual regulator

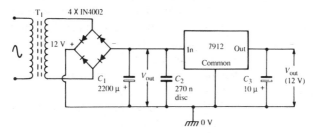

Figure 8.14 *Connections for using a 3-terminal negative regulator, in this case a 12 V 1 A '79' type*

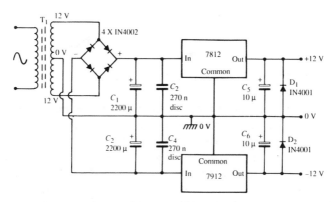

Figure 8.15 *Complete circuit of a 12 V 1 A dual power supply using 3-terminal regulator ICs*

circuits respectively. The ICs shown are 12 V types with 1 A ratings, but the basic circuits are valid for all other voltage values, provided that the unregulated input is at least 3 V greater than the desired output voltage. Note that a 270 nF or greater disc (ceramic) capacitor must be wired close to the IC's input terminal, and a 10 μF or greater electrolytic is wired across the output. The regulator ICs typically give about 60 dB of ripple rejection, so 1 V of input ripple appears as a mere 1 mV of ripple on the regulated output.

Voltage variation

The output voltage of a 3-terminal regulator IC is actually referenced to the ICs 'common' terminal, which is normally (but not necessarily) grounded. Most regulator ICs draw quiescent currents of only a few milliamperes, which flow to ground via this common terminal, and the IC's regulated output voltage can thus be raised above the designed value by biasing the common terminal with a suitable voltage, making it easy to obtain odd-ball output voltage values from these 'fixed voltage' regulators. *Figures 8.16* to *8.18* show three ways of achieving this.

In *Figure 8.16* the bias voltage is obtained by passing the ICs

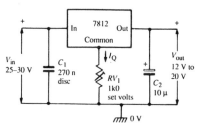

Figure 8.16   *Very simple method of varying the output voltage of a 3-terminal regulator*

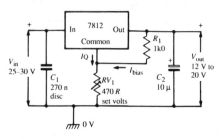

Figure 8.17   *An improved method of varying the output of a 3-terminal regulator*

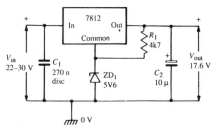

Figure 8.18 *The output voltage of a 3-terminal regulator can be increased by a fixed amount by wiring a suitable Zener diode in series with the common terminal*

quiescent current (typically about 8 mA) to ground via RV_1. This design is adequate for many applications, although the output voltage shifts slightly with changes in quiescent current. The effects of such changes can be minimized by using the *Figure 8.17* design, in which the RV_1 bias voltage is determined by the sum of the quiescent current and the bias current set by R_1 (12 mA in this example). If a fixed output with a value other than the designed voltage is required, it can be obtained by wiring a Zener diode in series with the common terminal as shown in *Figure 8.18*, the output voltage then being equal to the sum of the Zener and regulator voltages.

Current boosting

The output current capability of a 3-terminal regulator can be increased by using the circuit of *Figure 8.19*, in which current

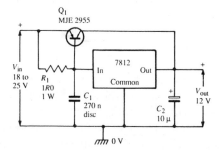

Figure 8.19 *The output current capacity of a 3-terminal regulator can be boosted via an external transistor. This circuit can supply 5 A at a regulated 12 V*

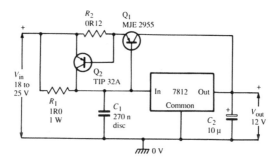

Figure 8.20   *This version of the 5 A regulator has overload protection provided via* Q_2

boosting can be obtained via bypass transistor Q_1. Note that R_1 is wired in series with the regulator IC; at low currents insufficient voltage is developed across R_1 to turn Q_1 on, so all the load current is provided by the IC. At currents of 600 mA or greater sufficient voltage (600 mV) is developed across R_1 to turn Q_1 on, so Q_1 provides all current in excess of 600 mA.

Figure 8.20 shows how the above circuit can be modified to provide the bypass transistor with overload current limiting via 0.12 Ω current-sensing resistor R_2 and turn-off transistor Q_2, which automatically limit the output current to about 5 A.

Variable 3-terminal regulator circuits

The 78xxx and 79xxx range of 3-terminal regulator ICs are designed for use in fixed-value output voltage applications, although their outputs can in fact be varied over limited ranges. If the reader needs regulated output voltages that are variable over very wide ranges, they can be obtained by using the 317K or 338K 3-terminal 'variable' regulator ICs.

Figure 8.21 shows the outline, basic data and the basic variable regulator circuit applicable to these two devices, which each have built-in fold-back current limiting and thermal protection and are housed in TO3 steel packages. The major difference between the devices is that the 317K has a 1.5 A current rating compared to the 5 A rating of the 338K. Major features of both devices are that their output terminals are always 1.25 V above their 'adjust' terminals,

Parameter	317 K	338 K
Input voltage range	4–40 V	4–40 V
Output voltage range	1.25–37 V	1.25–32 V
Output current rating	1.5 A	5 A
Line regulation	0.02%	0.02%
Load regulation	0.1%	0.1%
Ripple rejection	65 dB	60 dB

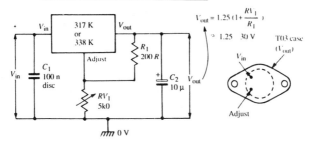

Figure 8.21 *Outline, basic data and application circuit of the 317K and 338K variable-voltage 3-terminal regulators*

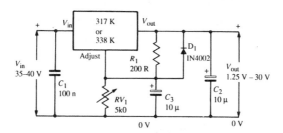

Figure 8.22 *This version of the variable-voltage regulator has 80 dB of ripple rejection*

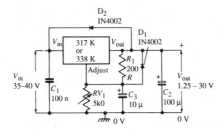

Figure 8.23 *This version of the regulator has 80 dB ripple rejection, a low impedance transient response, and full input and output short-circuit protection*

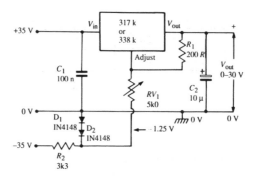

Figure 8.24   *The output of this version of the regulator is fully variable from zero to 30 V*

and their quiescent or adjust-terminal currents are a mere 50 μA or so.

Thus, in the *Figure 8.21* circuit, the 1.25 V difference between the 'adjust' and output terminals makes several milliamperes flow to ground via RV_1, thus causing a variable voltage to be developed across RV_1 and applied to the 'adjust' terminal. In practice the output of this circuit can be varied from 1.25 to 33 V via RV_1, provided that the unregulated input voltage is at least 3 V greater than the output. Alternative voltage ranges can be obtained by using other values of R_1 and/or RV_1, but for best stability the R_1 current should be at least 3.5 mA.

The basic *Figure 8.21* circuit can be usefully modified in a number of ways; its ripple rejection factor, for example, is about 65 dB, but this can be increased to 80 dB by wiring a 10 μF bypass capacitor across RV_1, as shown in *Figure 8.22*, together with a protection diode that stops the capacitor discharging into the IC if its output is short-circuited.

Figure 8.23 shows a further modification of the *Figure 8.22* circuit; here, the regulators transient output impedance is reduced by increasing the C_2 value to 100 μF and using diode D_2 to protect the IC against damage from the stored energy of this capacitor if an input short occurs.

Finally, *Figure 8.24* shows how the circuit can be modified so that its output is variable all the way down to zero volts, rather than to the 1.25 V of the earlier designs. This is achieved by using a 35 V negative rail and a pair of series-connected diodes that clamp the low end of RV_1 to -1.25 V.

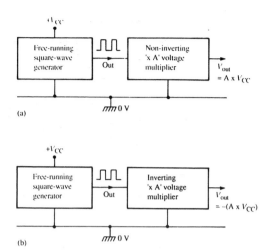

Figure 8.25 *(a) Voltage converter with positive output, and (b) voltage converter with negative output*

Voltage converter basics

Electronic voltage converters are circuits used to generate a higher-value supply from an existing low-voltage DC source, or to generate a negative DC supply from an existing positive DC voltage source, etc. Such circuits are fairly easy to design and build, using either readily available components or special-purpose IC's such as the ICL7660 voltage converter. A variety of low-power versions of such circuits are shown in the remainder of this chapter.

A DC voltage can easily be converted into one of greater value or reversed polarity by using the DC supply to power a free-running square-wave generator which has its output fed to a multisection capacitor-diode voltage multiplier network, which thus provides the desired 'converted' output voltage. If a positive output voltage is needed, the multiplier must give a non-inverting action, as in *Figure 8.25(a)*, and if a negative output is required it must give an inverting action, as in *Figure 8.25(b)*.

Practical converters of this type can use a variety of types of multivibrator circuit (bipolar or FET transistor, CMOS or TTL IC, etc.) as their basic free-running square-wave generators. In all cases, however, the generator should operate at a frequency in the range 1 kHz to 10 kHz, so that the multiplier section can operate

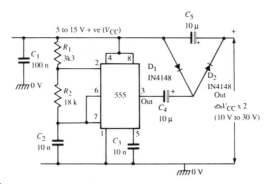

Figure 8.26   *DC voltage-doubler circuit*

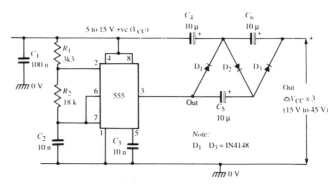

Figure 8.27   *DC voltage-tripler circuit*

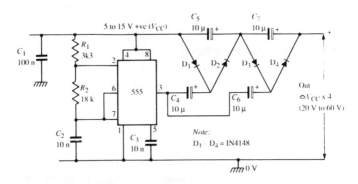

Figure 8.28   *DC voltage quadrupler circuit*

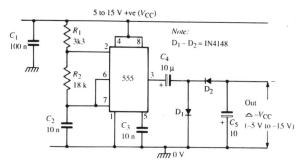

Figure 8.29 *DC negative-voltage generator*

with good efficiency while using fairly low values of 'multiplying' capacitor.

One of the easiest ways of making practical voltage converters of this type is to use type–555 'timer' ICs (which can supply fairly high output currents) as the free-running square-wave generators, and *Figures 8.26* to *8.29* show a selection of practical circuits of this type; in each case the 555 is wired as a free-running astable multivibrator and operates at about 3 kHz (determined by the R_1–R_2–C_2 values); C_1 and C_3 both help enhance circuit stability.

The *Figure 8.26* circuit acts as a DC voltage-doubler, and generates a DC output voltage roughly double that of the 555's supply line via the C_4–D_1–C_5–C_2 capacitor-diode voltage-doubler network, which produces the $2 \times V_{cc}$ output voltage.

This $2 \times V_{cc}$ is the approximate value of the unloaded output voltage; the precise value equals $2 \times V_{peak}, - (V_{df1} + V_{df2})$, where V_{peak} is the peak output voltage of the square wave generator and V_{df} is the forward volt drop (about 600 mV) of each 'multiplier' diode. The output voltage decreases when the output is loaded.

The *Figure 8.26* circuit can be used with any DC supply in the range 5 to 15 V and can thus provide a 'voltage doubled' output of 10 to 30 V. Greater outputs can be obtained by adding more multiplier stages to the circuit. *Figure 8.27*, for example, shows how to make a DC voltage tripler, which can provide outputs in the range 15 to 45 V, and *Figure 8.28* shows a DC voltage quadruple, which gives outputs in the 20 to 60 V range.

A particularly useful type of 555 converter is the DC negative-voltage generator, which produces an output voltage almost equal to amplitude but opposite in polarity to that of the IC supply line.

This type of circuit can be used to provide a split-supply output for powering op-amps, etc., from a single-ended power supply. *Figure 8.29* shows an example of such a circuit, which operates at 3 kHz and drives a voltage-doubler (C_4–D_1–C_5–D_2) output stage.

High-voltage generation

The 'voltage multiplier' method of generating increased values of output voltage is usually cost-effective only when multiplier ratios of less than six are needed. In cases where very large step-up ratios are required (as, for example, when hundreds of volts must be generated via a 12 V supply), it is often better to use the output of a low-voltage oscillator or square-wave generator to drive a step-up voltage transformer, which then provides the required high-value voltage (in AC form) on its secondary (output) winding; this AC voltage can easily be converted back to DC via a simple rectifier-filter network, if required. *Figures 8.30* to *8.32* show some practical low-power high-voltage generator circuits of these types.

The *Figure 8.30* circuit acts as a DC-to-DC converter which generates a 300 V DC output from a 9 V DC power supply. In this case Q_1 and its associated circuitry act as a Hartley L–C oscillator, with the low-voltage primary winding of 9 V–0–9 V to 250 V mains transformer T_1 forming the 'L' part of the oscillator, which is tuned via C_2. The supply voltage is stepped up to about 350 V peak at T_1 secondary, and is half-wave rectified and smoothed via D_1–C_3. With no permanent load on C_3, the capacitor can deliver a powerful but non-lethal 'belt'. With a permanent load on the output, the output falls to about 300 V at a load current of a few milliamperes.

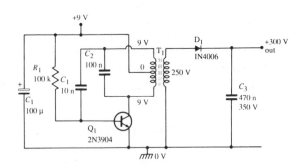

Figure 8.30   *9 V to 300 V DC-to-DC converter*

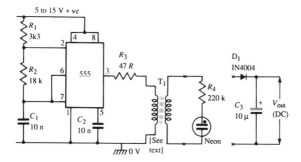

Figure 8.31 *Neon-lamp driver or 'high-voltage' generator*

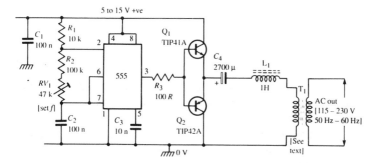

Figure 8.32 *DC-to-AC inverter*

The *Figure 8.31* circuit can be used to drive a neon lamp or generate a low-current high-value (up to a few hundred volts) DC voltage from a low-value (5 to 15 V) DC supply. The 555 is wired as a 3 kHz astable multivibrator that drives T_1 via R_3. T_1 is a small audio transformer with a turns ratio sufficient to give the desired output voltage, e.g., with a 10 V supply and a 1:20 T_1 turns ratio, the transformer will give an unloaded DC output of 200 V peak. This AC voltage can easily be converted to DC via a half-wave rectifier and filter capacitor, as shown.

Finally, the *Figure 8.32* DC-to-AC inverter circuit produces an AC output at mains power-line frequency and voltage. The 555 is wired as a low-frequency (variable from 50 to 60 Hz via RV_1) astable that feeds its power-boosted (via Q_1–Q_2) output into the low-voltage 'input' of reverse-connected filament transformer T_1, which has the desired 'step-up' turns ratio. C_4 and L_1 act as a filter that ensures that the power signal feeding into the transformer is essentially a sine wave.

The ICL7660

The ICL7660 is a dedicated voltage converter IC specially designed to generate an equal-value negative supply from a positive source, i.e., if powered from a +5 V supply it generates a −5 V output. It can be used with any +1.5 to 10 V DC supply, and has a typical voltage conversion efficiency of 99.9 per cent (!) when its output is unloaded; when the output is loaded it acts like a voltage source with a 70 Ω output impedance, and can supply maximum currents of about 40 mA.

The ICL7660 is housed in an 8-pin DIL package as shown in *Figure 8.33*. The IC actually operates in a way similar to that of the *Figure 8.26* 'oscillator and voltage-multiplier' circuit, but with far greater efficiency. The ICL7660 chip houses a very efficient square-wave generator that operates (without the use of external components) at about 10 kHz and has an output that switches fully between the supply rail values. It also houses an ultra-efficient set of logic-driven multiplier 'diodes' that, when used with two external capacitors, enables voltage-doubling to be achieved with near-perfect efficiency.

The reader may recall from the *Figure 8.26* description that the use of conventional multiplier diodes makes the circuit's unloaded output drop by about 1.2 V, this being the sum of the forward volt drops of the two diodes. In the ICL7660 this volt drop is eliminated by replacing the diodes with MOS power switches, driven via logic networks in such a way that each 'diode' switch automatically closes when it is forward biased and opens when reverse biased, thus giving near-perfect operating efficiency.

The ICL7660 is an easy device to use, but note that none of its terminals must ever be connected to a voltage greater than V+ or

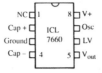

Figure 8.33   *Outline and pin notations of the ICL7660 voltage converter IC*

less than *ground*. If the IC is to be used with supplies in the range 1.5 to 3.5 V, the pin-6 'LV' terminal should be grounded; at supply values greater than 3.5 V, pin-6 must be left open circuit. At supply values greater than 6.5 V a protection diode must be wired in series with *output* pin-5. *Figures 8.34* to *8.42* show a selection of practical application circuits in which these design rules are applied.

ICL7660 circuits

The most popular application of the ICL7660 is as a simple negative voltage converter, and *Figures 8.34* to *8.36* show three basic circuits of this type. In each case, C_1 and C_2 are 'multiplier' capacitors and each have a value of 10 μF.

The *Figure 8.34* voltage converter is intended for use with 1.5 to 3.5 V supplies, and requires the use of only two external components. The *Figure 8.35* circuit is similar, but is meant for use with supplies in the 3.5 to 6.5 V range and thus has pin-6 grounded.

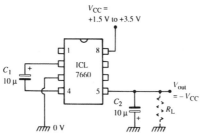

Figure 8.34   *Negative voltage converter using 1.5 V to 3.5 V supply*

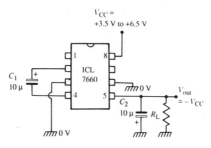

Figure 8.35   *Negative voltage converter using 3.5 V to 6.5 V supply*

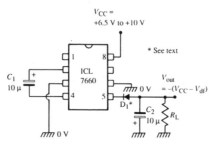

Figure 8.36   *Negative voltage converter using 6.5 V to 10 V supply*

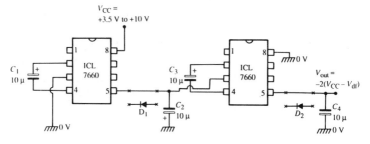

Figure 8.37   *Cascading devices for increased output voltage*

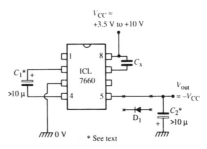

Figure 8.38   *Reducing oscillator frequency*

Finally, the *Figure 8.36* circuit is meant for use with supplies in the range 6.5 to 10 V, and thus has diode D_1 wired in series with output pin-5, to protect it against excessive reverse biasing from C_2 when the power supplies are removed. The presence of this diode reduces the available output voltage by V_{df}, the forward volt drop of the diode; to keep this volt drop to minimum values, D_1 should be a germanium type.

A useful feature of the ICL7660 is that numbers of these ICs (up to a maximum of ten) can be cascaded to give voltage conversion factors greater than unity. Thus, if three stages are cascaded, they give a final output voltage of $-3\ V_{cc}$, etc. *Figure 8.37* shows the connections for cascading two of these stages; any additional stages should be connected in the same way as the right-hand IC of this diagram.

In some applications the user may want to reduce the ICL7660's oscillator frequency; one way of doing this is to wire capacitor Cx between pins 7 and 8, as in *Figure 8.38*. *Figure 8.39* shows the relationship between the C_x and frequency values; thus a C_x value of 100 pF reduces the frequency by a factor of ten, from 10 kHz to 1 kHz. Note that, to compensate for this 10:1 frequency reduction

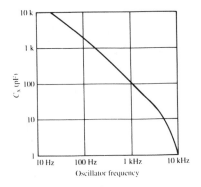

Figure 8.39   *C_x/oscillator frequency graph*

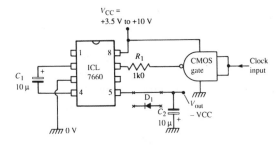

Figure 8.40   *External clocking of the ICL7660*

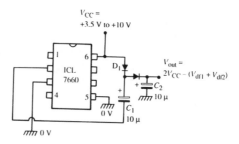

Figure 8.41   *Positive voltage multiplier*

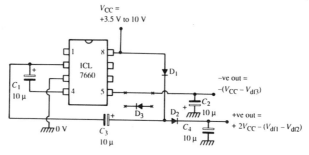

Figure 8.42   *Combined positive voltage multiplier and negative voltage converter*

and maintain the circuit efficiency, the C_1 and C_2 values should be increased by a similar factor (to about 100 μF each).

Another way of reducing the oscillator frequency is to use pin-7 to over-drive the oscillator via an external clock, as shown in *Figure 8.40*. The clock signal must be fed to pin-7 via a 1k0 series resistor (R_1), and should switch fully between the two supply rail values; in the diagram, a CMOS gate is wired as an inverting buffer stage, to ensure such switching.

Another use of the ICL7660 IC is as a positive voltage multiplier, to give a positive output of almost double the original supply voltage value. *Figure 8.41* shows the circuit connections. The pin-2 oscillator output signal is used here to drive a conventional capacitor-diode voltage-doubler network, of the type used in *Figure 8.26*. Note that these two diodes reduce the available output voltage by an amount equal to their combined forward volt drops, so they should ideally be low-loss germanium types.

Finally, to complete this look at ICL7660 applications, *Figure*

8.42 shows how the circuits of *Figure 8.35*, *8.36* and *8.41* can be used to make a combined positive voltage multiplier and negative voltage converter that provides dual output voltage rails. Each rail has an output impedance of about 100 Ω.

Index

Types of integrated circuit (IC)